ENTREPRENEURIAL ETHICS AND TRUST

The Making of Modern Africa

Series Editors: Abebe Zegeye and John Higginson

Growth or Stagnation? South Africa heading for the year 2000
Mats Lundahl

Sudan's Predicament
Civil war, displacement and ecological degradation
Edited by Girma Kebbede

Doctors and the State
The struggle for professional control in Zimbabwe
Dorothy Mutizwa-Mangiza

Class Formation and Civil Society: The Politics of Education in Africa
Patrick M. Boyle

The Social Services Crisis of the 1990's
Edited by Anna Kajumulo Tibaijuka

Oil and Fiscal Federalism in Nigeria
Augustine A. Ikein and Comfort Briggs-Anigboh

Environment, Health and Population Displacement
Andrew E. Collins

International Banking and Rural Development
Pade Badru

Structural Adjustment and Mass Poverty in Ghana
Kwabena Donkor

Contemporary Issues in Socio-economic Reform in Zambia
Edited by Herrick C. Mpuku and Ivan Zyuulu

The State and Organised Labour in Botswana
Monageng Mogalakwe

Entrepreneurial Ethics and Trust

Cultural foundations and networks in the Nigerian plastic industry

YAKUBU ZAKARIA

Ashgate

Aldershot • Brookfield USA • Singapore • Sydney

Published by
Ashgate Publishing Ltd
Gower House
Croft Road
Aldershot
Hants GU11 3HR
England

Ashgate Publishing Company
Old Post Road
Brookfield
Vermont 05036
USA

Ashgate website: http://www.ashgate.com

British Library Cataloguing in Publication Data
Zakaria, Yakubu
Entrepreneurial ethics and trust: cultural foundations and networks in the Nigerian plastic industry. - (The making of modern Africa)
1. Plastics industry and trade - Nigeria 2. Business ethics - Nigeria
I. Title
338.4'76684'09669

Library of Congress Catalog Card Number: 99-73623

ISBN 0 7546 1160 4

Printed in Great Britain

Contents

Part II: The Nigerian Plastic Industry, Entrepreneurial Profit Motive, and Resource Controls

Part III: General Summary, and Conclusions

List of Tables, Figures and Maps

Tables

Figures

Maps

Abbreviations

ACCs:	Advanced Capitalist Countries
AD:	Anno Domini (year of the Christian Calendar)
AFEM:	Autonomous Foreign Exchange Market
AFF:	Advanced Fee Fraud (419)
AH:	Anno Hegirae (year of the Hijra)
BOP:	Balance of Payment
CAC:	Corporate Affairs Commission
CBN:	Central Bank of Nigeria
CFAO:	Compagnie Francais D'Afrique Occidentale
COP:	Current Operating Profit
FAO:	Food and Agricultural Organisation of the United Nations
EIU:	The Economist Intelligence Unit
FEPA:	Federal Environmental Protection Agency
FTSE:	Financial Times Stock Exchange
IAP:	Industrial Arbitration Panel
ILO:	International Labour Organisation
ITF:	Industrial Training Fund
KSG:	Kano State Government
LDCs:	Less Developed Countries
MAN:	Manufactures' Association of Nigeria
NECA:	Nigerian Employers Consultative Association
NEPA:	Nigerian Electric Power Authority
NEPD:	Nigerian Enterprises Promotion Decree
NEPU:	Northern Elements Progressive Union
NIC:	National Industrial Court
NPF:	National Provident Fund
NLC:	Nigerian Labour Congress
NNPC:	Nigerian National Petroleum Corporation
NPC:	National Population Commission
OPEC:	Organisation of Petroleum Producing Countries
PPIB:	Price Productivity and Incomes Board

PRP:	People's Redemption Party
PVC:	Polyvinylchloride
SAP:	Structural Adjustment Program
SCOA:	Societé Commerciale Ouest Africaine
SON:	Standards Organisation of Nigeria
UAC:	United African Company
UNCTAD:	United Nations Conference on Trade and Development
UNESCO:	United Nations Educational, Scientific, and Cultural Organisation

Preface and Acknowledgements

My aim in this book is to provide a comprehensive analysis of how culture influences entrepreneurial ethics and trust in a northern Nigerian society centred in Kano. I begin with the assumption that culture does not operate in a vacuum. It influence political, economic, and social institutions and vice versa.

The book is not intended to explore all the broad aspects of culture and societal development. Rather, it is set to examine the interplay among culture, entrepreneurial ethic and trust relations in the northern Nigerian plastic industry. To provide a meaningful interpretation of the cases, various cross-cultural paradigms, and their relevance for providing an understanding of the culturally peculiar patterns in the business organisations are reviewed. Apart from the plastic sector, the book explores wide-ranging topics that strike the core of contemporary business behaviour in Nigeria and many developing societies.

To arrive at a meaningful understanding of contemporary business behaviour, the historical evolution of indigenous entrepreneurship is examined. A look at the link between culture and entrepreneurial rationality is significant to the study. Thus, the logics between religious reform and new forms of entrepreneurial rationalisations are crucial in this regard. The analysis of rationality in Hausa society is extended in the latter part of the study to cover hidden economic activities of secluded women.

Entrepreneurial trust relations and their implications for the business sector are crucial to this study. In the concluding sections of the study, an attempt is made at reformulating the modernisation theory from a cultural perspective.

Having said this, it will be most remiss on my part not to mention the important contributions of people and institutions from which various kinds of support was received. First and foremost, my sincere thanks to Professor Emeritus Ulf Himmelstrand for reading early drafts of this book and providing the necessary guidance. The association with Professor Himmelstrand has yielded immense academic profit and highly cherished personal experience in terms of patience. To Professor Tom R. Burns, Dr. Endre Stiansen and Steve Turner, I am most indebted for the incisive and constructive criticisms on the early drafts of the manuscript. Professors Peter Ekeh and Said Arjomand were very effective in drawing my attention to new topics and relevant literature on the subject. In preparation of this book, I have also benefitted from the editorial talents of Leslie Irvine, Minna Richt and Emily Merson. None of these people should be held responsible for any of the views I express.

Special thanks are due to the following libraries for permitting me to use their resources: the Kano State Library, the British Council Library in Kano, the Uppsala University Library, the Swedish Parliament Library in Stockholm, the Nordiska African Institute Library, the Dag Hammarskjöld Bibliotek, the School of Oriental and African Studies Library in London, the Cambridge University Library, the University of Gothenburg Library, the University of Lund Library, the Stockholm University Library, and the Departments of Anthropology, Theology, Business Studies and Economics at Uppsala University, Sweden.

I am deeply grateful to Alhaji Kabiru Yakasai, head of the Inspectorate Division of the Kano State Ministry of Commerce and Industry, for providing the link with the plastic manufacturers of Kano. To the indigenous and expatriate entrepreneurs, managers and employees of the plastic factories, whom I promised anonymity, I wish to express my sincere thanks for granting me access to their factories.

I wish to record my gratitude to Eva Gustafsson of Hancock and Rehn Chambers, Uppsala, whose support at the early and critical stages of this work has been very helpful. In particular, I want to thank the families of Professor Adebayo Olukoshi, Dr. Desmond A. Aboagye, Dr. Fred O. Asiegbu and Nicholas Baafi, for providing the cherished friendship and emotional support during these years. To the members of my family, and especially, Is-haq, Rabi, Samira, and Farida, I wish to express my sincere appreciation for continued understanding and support of my academic career.

Part I
Method of Analysis, and Theoretical Orientation

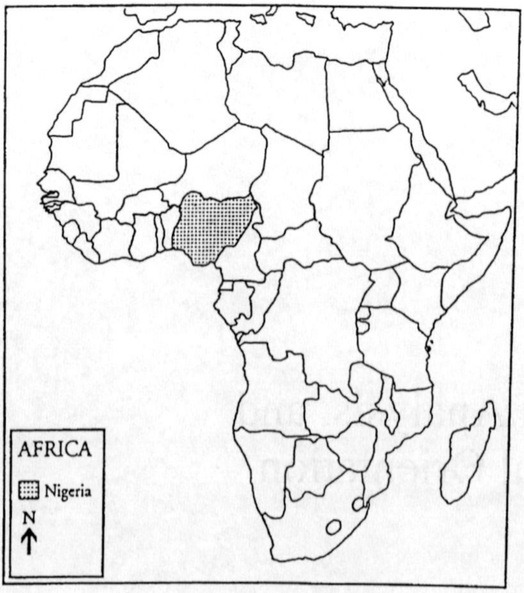

Map 1.1. Nigeria in Africa
***Source:* Survey Department Kano.**

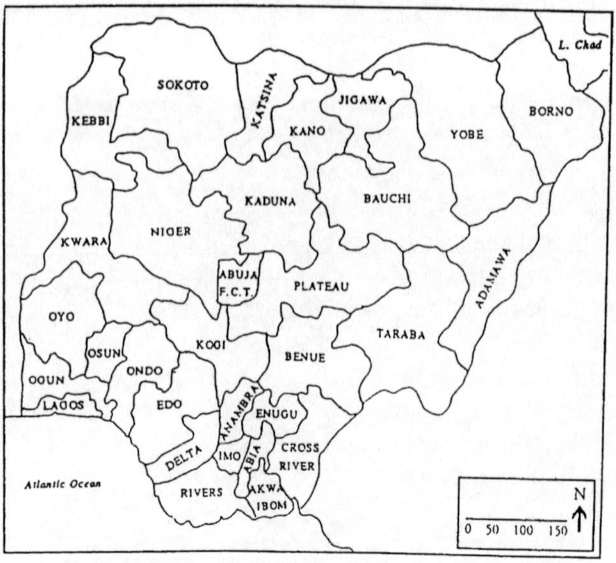

Map 1.2. Nigeria: State Boundaries, 1993–1996
***Source:* Survey Department Kano.**

1 Introduction

> Cultural analysis is intrinsically incomplete. And, worse than that, the more deeply it goes the less complete it is. It is a strange science whose most telling assertions are its most strenuously based, in which to get somewhere with the matter at hand is to intensify the suspicion, both your own and that of others, that you are not quite getting it right (Geertz, 1993, p. 29).

Culture is relevant. It is relevant not just because it is a repository of societal values, habits and conventions. It also provides rules and beliefs for what individuals and groups do. Societal values, rules, and belief systems pave the way for rationality, which provides guidance on how individuals and groups pursue economic development. Culture's significance lies in the sense that it regulates the rate at which societies and groups adapt to change. Despite the 'aura of ill-repute', 'multiple referents', and 'vagueness'[1] surrounding cultural explanations, they are useful in providing us with the subtle meanings that underlie ethics and interpretations about what is rational or irrational economic behaviour across societies. Yet, culture does not operate in a vacuum. Of course, it influences political, economic, and social institutions in society and vice versa.

The study features three firms with distinct cultures and subcultures. Central to the outline of empirical analysis is the cultural dimension of economic performance in the northern Nigerian plastic industry. Of particular significance to the study is the proposition that cultures vary in the way they influence economic performance across societies. Much of the analysis in this book centres on how cultural and religious[2] values influence entrepreneurial ethics and trust relations in business transaction. The central focus and important question in this triad study is: how does culture influence entrepreneurial ethics and trust in a developing society?

There is no single but a variety of cultural responses to modernity across societies. As such, when capitalism penetrates developing societies, indigenous cultural values usually determine the social formations of production. Capitalism often assumes a syncretic form peculiar to the cultural environment.[3] Culture is therefore not arcane but germane to the study of societal development process.

Equally relevant to this study is the social transformation involved in the adoption of new patterns of manufacturing activities in northern Nigeria where there is much of a heritage in commerce to build on. Today, manufacture involves the introduction of new technologies and social patterns. From a sociological point of view, the most significant aspect of this process is the interplay among these new economic activities and the indigenous patterns of culture and religion.

Culture, a key concept in this study, is an umbrella-term that refers to the all-encompassing material and non-material shared aspect of the human heritage, which includes values, norms, customs, and belief systems that are governed by rules. Culture has three dimensions: the cognitive, normative, and conative. These dimensions, which correspond to thinking, judging, and acting, are regulated by established rules within a particular group. Thus, culture is applied in a pluralistic sense to recognise the existence of subcultural groups within the northern Nigerian society and the business organisations under study. Cultural and subcultural groups often have values, norms and myths that may be similar or distinct from each other.

Values, on the other hand, are 'relatively general and durable internal criteria for evaluation' (Hechter, 1993, p. 3). Although they differ in scope and application, they largely influence people's subjective definitions of what constitutes rationality and which outcomes are labelled as good or bad. Instrumental[4] values exercise profound influence on trust, risk, and investment attitudes. Human action depends on internalised values that are acquired or learned through a process of socialisation. Social action and verbal statements provide cues to personal values (Kunkel, 1971). On a general level, societal values and norms tend to exert influence on the individual. This, however, does not foreclose personal or individual values, which sometimes do clash with the values of the larger society. Since values are interwoven with past and present experiences, previous disappointments with old values make people receptive to new ones. Unlike values, norms are evaluative, durable, and external to actors. Norms require sanctions for their efficacy (Hechter, 1993, pp. 2–28).

Islamic and Hausa cultural values are regarded as influences quintessential to understanding entrepreneurial ethics in northern Nigeria. In order to describe these tendencies, entrepreneurial ethics are characterised as both a 'set of normative prescriptions', 'emotions' and 'dispositions',[5] which underlie business behaviour. Business, in this context, refers to the sum of organised or patterned activities by which people engage in commerce and industry to provide goods and services and for the acquisition of income.

Business is conceptualised in a much broader context than in its daily Hausa usage. In Hausa society, business activity has several linguistic equivalents. These include *koli*, which implies the peddling of small wares; *bojuwa*, or long distance commercial activity; *ciniki* and *kasuwanci,* which translate as acts of trade; and *sana'a,* or occupation. This implies that itinerant vendors and food hawkers with very little capital are regarded as businesspeople. They are considered the same as those engaged in large commercial and manufacturing activities with substantial capital at their disposal. In attempting to overcome these ambiguities, the business activity on which the study is focused is the plastic manufacturing industry in northern Nigeria. In order to grasp the nettle firmly, entrepreneurial strategies in the modern manufacturing activities are examined from cultural and historical perspectives.

Distinct Characteristics of the Study

The distinct characteristics of the study are as follows: First, it investigates the role of Hausa and Islamic cultural values in shaping specific entrepreneurial ethics in northern Nigeria. Essentially, the study provides a cross-cultural comparison of Hausa, Lebanese, and Chinese plastic firms in the northern Nigerian city of Kano. Such a comparative analysis on how local cultural values penetrate modern business organisations is rare. For many years, Nigerian social scientists have underestimated the role of culture in economic development. Although both indigenous and foreign scholars have studied local businesses, few have investigated the cultural impact on entrepreneurial ethics and trust patterns. The plethora of macro-studies, which often do not reveal micro-empirical links, are based on tacit assumptions that economic and political factors operate independently of societal values. This study attempts to bridge the lacuna in the Nigerian business literature.

In spite of the significant cultural and economic differences between especially the Hausa and Chinese, such a comparison makes sense because both groups represent cultures with a strong sense of family. My field studies of 1994/95 and a subsequent follow-up in 1998, shows that most Hausa, Chinese, and Lebanese economic activities are labour-intensive family businesses whose operations are sometimes shrouded in secrecy. In particular, entrepreneurial investment patterns are influenced by a form of rationality that suggests low degree of trust in the business environment. Trust is crucial in influencing investment strategies in Nigeria as in many developing countries. The book explores contributions of private entrepreneurs in the socioeconomic transformation of northern Nigeria, and challenges Weber's viewpoint that economic progress is less likely in an Islamic society.

A fourth distinct feature of this study is the focus on entrepreneurial performance in the modern manufacturing sector[6] from a cultural perspective. The evidence in Southeast Asia suggests that there is no single path to economic development. This shows that genuine development processes need to be approached from cultural vantage points. Thus, sustainable development must consider peculiar local circumstances. Cultural studies are significant in the sense that they allow us to capture the indigenous point of view on how economic goals could be pursued under a given situation. In a semi-industrial city like Kano, business operators and factory workers live in two worlds. In one world, they imbibe the cultural and religious values of the general society; in the other, they are engaged in opposing forms of modern economic activity involving foreign technology and capitalist modes of production.[7] An examination of how these dual roles are encountered and reconciled in the economic process is significant. It may suffice to mention therefore, that this comparative study of indigenous and expatriate firms, via cultural processes, is justifiable.

Following logically from these queries and possible conjectures, this study has three main objectives. Firstly, to do a cross-cultural examination of entrepreneurial and managerial strategies. Secondly, to explore entrepreneurial trust and risk-related attitudes that exert significant influence on the nature and size of business enterprises. Here, the patterns and structure of business and their possible effects on entrepreneurial resource controls in indigenous and expatriate firms are relevant. Finally, the study seeks to provide an understanding of the nature and impact of cultural values on the three plastic firms.

Method, and Unit of Analysis

The research reported here makes use of a variety of methods which include: interviews, observations, analysed records from three firms, and reanalysis of secondary materials particularly concerning labour market and entrepreneurial activity from historical perspectives. The units of analyses are three business firms in the plastic sector of Kano, which include an indigenous Hausa, a Lebanese, and a Chinese firm. In addition, some major activities in the commercial sector and their relevance on the modern industrial sector of Kano are examined. The selection procedure of firms was based on accessibility.

There are currently over 60 plastic firms operating within the Kano metropolis. Three firms in the plastic sector, with a total population of 820 employees from which a random sample (N=180), was drawn, form the basis of the interview population. Respondents in the study consist of 14 managers and 166 employees in the three plastic manufacturing firms. In addition, 30 interviews were conducted among artisans, and traders outside the three firms. Together, 210 respondents were interviewed.

These respondents, with the proprietors of the plastic firms, are labelled as entrepreneurs in the entire study. Although entrepreneurs are treated as distinct from managers, entrepreneurial, and managerial responses are often lumped together in this study to distinguish them from lower-rank factory employees. Additional information was also received from important figures in the Kano State Ministries of Industry and Commerce. Although the sampling frame may not be strictly representative of a specified population, Zetterberg's (1965, pp. 128-130) study shows that representativeness is not the only goal in sampling, variation or scope is another. The use of three case studies with considerable variations in cultural and economic potentials is therefore considered adequate. Case studies are useful in providing an understanding of areas of enterprise or organisational function that is often not well documented or amenable to investigation by other methods.[8]

Entrepreneurial performance regarding modern manufacturing activities is examined, with particular interest paid to the peculiar character of each firm. The approach in this study is comparative and intra-organisational in scope. Essentially, three case studies are compared. The study is largely a fusion of quantitative and qualitative methods. Multiple methods were adopted to tap and maximise the strength of each method. This allows the reduction and prevention of total error through cross-checking and comparing the findings elicited by each method. Web *et al.* (1966) stress the relevance of combining more than one method in research investigation for providing reliable results.

The focused interview method was considered as the most appropriate investigation tool for this study. The interviews focused on variables considered important in analysing entrepreneurial behaviour. In addition, single and group interviews were conducted with key businesspeople and workers in each firm. This approach changed when I interviewed factory workers on the shop floor. Because of their large numbers, group interviews were conducted in which up to a maximum number of ten workers were sometimes randomly selected and interviewed together. Duration of the interviews varied from one hour to one and one-half hours, depending on the setting.

The focused interview method has the ability to reveal 'deep-seated trends or values' of the persons being interviewed. Merton, *et al.* (1956), first popularised the method. Its distinct characteristics are as follows. First, the persons interviewed are known to have been involved in some specific kind of activity. Second, a 'content' or 'situation' is made of the activities in which the persons are engaged. Third, the 'situational' or 'content' examination provides an interview guide for other related areas of enquiry. Finally, interviews focus on 'the subjective definitions of persons exposed to the preanalysed situation'. Attention is therefore given to the retrospective and introspective experiences of the interviewed persons. The special application of the focused interview was derived from the economic activities of the entrepreneurs in the plastic-manufacturing sector. Identification of entrepreneurs and businessmen was derived from type of economic activity, tangible and physical assets such as buildings, machines, or the workers they employed. Interview questions focused on the subjective definitions of entrepreneurs on risk, profit, business success and the economic situations or challenges they faced.

In designing the interview questions, the 'funnel technique' was applied. In this technique, interview questions normally begin from the general and move towards the specifics. Its importance lies in the clarification of general questions. Since most of the questions were open-ended, a great amount of care was needed to prevent the provision of 'single links' or responses by adopting a technique of 'asking why' (Lazarsfeld, 1935).[9] In-depth interviews were conducted with entrepreneurs, managers, and workers who played essential roles in the firms. All the recorded interviews were later transcribed verbatim on paper, coded, and processed through the statistical method.

Some advantages of the interview method are that literacy is not necessary, especially if the interviewer understands and speaks the local language. This personal approach allows for intimate knowledge and behaviour of respondents (Peil, 1982). An additional advantage of the focused interview is its ability to yield both anticipated and unanticipated results within a short period. It allows for a range of both intensive and extensive interview questions. Furthermore, it enables the interviewer to pick up important cues, from which additional questions and information can be obtained. Some disadvantages of the method, however, are the tendency to put the interviewer in the position of 'arresting comments' or 'cutting leads open before they are fully developed'. There is also a tendency to fixity and the imposition of a fallacy of preconceived topics which respondents may not be familiar with or even exposed to (ibid.).

Participant observation[10] in this study is a combination of obtrusive and unobtrusive methods. A major advantage of this approach is the tapping of additional information which respondents could not provide in the interview process. Three informants were hired in the factories for acquiring additional information which the management and owners of the firms may not have provided. The entire period of the field study was six months. During the fieldwork, I observed and compared specific nonverbal behaviour with verbal utterances of both entrepreneurs and employees for careful documentation. A basic advantage of this method is that it provides information on events directly as they occur in their natural setting.

Sources of secondary materials[11] include historical records, factory documents, and relevant information from the Federal and State Ministries of commerce and industry in Kano. A study of entrepreneurial patterns from a cultural perspective requires an examination of archival and historical materials. This is because 'history and sociology are inseparably linked' (Calhoun, 1995, p. 141). The historical approach in cultural studies exposes a variety of otherwise uncontrollable variables which cannot be directly approached by purely empirical and theoretical analyses (Cochran, 1971). In this regard, answers may only come through 'analogies of historical situations'. The importance of historical material is that data can cover a much longer time span than is feasible with other approaches of data collection.[12]

Through a study of company records, cultural factors that impinge on organisational behaviour can be elucidated. Company records are also important in the sense that sometimes information about inaccessible persons like business executives and shareholders can be easily obtained. Such records provide insight into history's underlying processes and the changes that appear under different epochs within a firm. Thus, secondary materials are very useful in providing an understanding of the peculiarities in organisations.

Cultural Analyses of Economic Development

For many years, cultural explanations of human performance have generated controversies in academic circles. Scholars in almost all academic disciplines have abused discourses with cultural links to particularities and especially economic development. In anthropology, the 'fear of historicism' and the prolonged occupation with the idea of 'cultural relativism' was a major factor that hindered the development of such analysis (Geertz, 1993, pp. 43–44). Ironically, the biological school, for instance, saw certain cultures 'as pathetically inept', possessing neither 'enterprise' nor 'intellect' (Frederickson, 1971, p. 106). Over the years, cultural labels fed stereotypes, endorsed static and uniform views, and even promoted segregation (Klitgaard, 1992, p. 107). Invariably, human character and ability have become associated with cultural traits. Cultural concepts have been perceived over the years as racists' analytical tools. In view of this, the prejudices against cultural explanations in contemporary development research have been strong.

During the early years of independence, many scholars and political leaders in young African states such as Nigeria, Ghana, and Gambia shied away from cultural discourses relating to development, this depicted indigenous cultures as inferior. The early political rhetoric about self-reliance therefore did not incorporate cultural views. Paradoxically, culture became a much-abused term among African political leaders in the post-independence era. In the political sphere, culture has become a rhetorical conundrum and justification for the one-party state and human right abuses in most African states. The false reference to culture often becomes 'an alibi for evading the political problem or national liberation'. Under such circumstances, the 'hypertrophy of cultural nationalism generally serves to compensate for the hypertrophy of political commitment' (Hountondji, 1983, p. 159).[13]

On the premise that the 'seeds of culture' are antithetical to modernisation, African scholars have avoided cultural studies on economic development. The implicit notion in this assumption is the conception of culture as static. Available evidence, however, suggests the contrary. It has been observed that 'sometimes after long periods of stagnation in a society, there emerges a new found capacity to absorb the new and reinvent the past, as China has done twice in half a century' (Serageldin, 1992, p. 1). Furthermore, recent studies on development have weakened the hypothesis that modernisation inherently contains the seeds of cultural destruction. Salim (1992, p. 12) observes that the widely held view that culture has within it the seed of its own destruction in modernisation is untrue as suggested in Japan and Southeast Asia.

Presumably, cultural studies have been unpopular especially among scholars for the fact that culture is a complex and loaded term that many social scientists prefer to ignore. Empirical studies featuring culture as an outcome or as a causal variable are considered highly controversial. At the descriptive and exploratory levels, the notion of culture remains vague (Archer, 1989, p. 1). Until recently, cultural explanations have generally been regarded as weak because they usually describe or point to issues that can hardly be addressed or answered by development policies.

The neglect of culture and micro-institutional structures has led to a scenario where development management in most African communities has become solely a government domain. Experiences in Kano and parts of Northern Nigeria have shown that communal participation in development efforts is often negligible, except where the local functionary, such as king or *sarki*, and ward head or *mai-unguwa*, was directly involved. This underscores the relevance of traditional institutions in development effort.[14] In particular, the knowledge of what, how, where, and when to do things is relevant to development planning and begins with an understanding of indigenous cultures (Klitgaard, 1992). However, it is necessary to mention that cultural factors alone do not determine economic performance or development. Culture does not operate in a vacuum, it interacts with other variables, and they all influence each other. However, since development is generally aimed at improving the quality of life in a society or country, its links with culture are naturally inevitable.

Culture's Relevance

There is now a growing interest about the role of culture in development.[15] This renewed interest has led to the introduction of cultural development-related programmes under the aegis of international organisations such as the World Bank and UNESCO. The cultural dimension of development has now become one of the most relevant approaches through which donor agencies can obtain adequate information and understanding of the problems in developing countries. Not surprisingly, UNESCO has labelled the 1990s the World Decade for Cultural Development. Until now, the dominant political-economic approach has always addressed Africa's problems from a 'top-down' perspective.

The relevance of the cultural approach to development is that it emphasises a 'bottom-up' or empowerment strategy that enables a group to affirm itself in meaning and in self-confidence (Klitgaard, 1992, p. 111). Conversely, imposed development often leads to cultural resistance (Bryant, 1992). The vibrancy[16] of the informal sector in Africa, in the midst of generalised economic crises and difficulties of the organised sector illustrate the limits of any linear conception. For Dia (1992, p. 172), the success of most of these micro-enterprises can be explained by their ability to reconcile African social and cultural values with the need for economic efficiency. Culture is about management because management concerns institutional analysis, motivation, establishing incentives exercising authority, mobilising and leading people and carrying out decisions (Bryant, 1992).

Cultural studies are relevant because normally, they point to the internal factors responsible for the perpetuation of the current economic imperfections. This invariably provides clues on how local obstacles can be minimised or eliminated. Notwithstanding the external constraints on Africa's economic development, the role of the 'environment' and of the 'super-structure of the human society' in shaping the patterns and nature of development, cannot be entirely ignored (Rodney, 1972, p. 16).[17]

Countries such as Japan, South Korea, and Taiwan, have successfully modernised without renouncing local customs. Through modernisation and economic advancements, the countries in Southeast Asia have proven that tradition and modern technological advancements can coexist (Dia, 1992, p. 180). That Kano has lost its pre-colonial importance as a textile manufacturing centre in sub-Saharan Africa because it could not expand the sector into a modern industry shows the negative consequences of a neglect of indigenous cultural values.

Modern technology provides the basic instruments that can improve productivity and general living standards. The nontechnical colonial educational legacy inherited by most African countries is faulty and incapable of propelling modern development. A proper understanding of indigenous cultures is a basic requirement for discovering the most appropriate educational system for a country's development. However, since technology is not culture-neutral, cultural receptacles should be adapted to the technical instruments. The history of modern technology reveals that nearly every new invention is the product of a previous crude invention elsewhere. Poor and ill-adaptive educational systems make technological transfers unattainable. Indigenous Nigerian cultures must learn to incorporate the language of modern technology so that technical instruments can become manipulable tools through innovation, imitation, and duplication. Today, the 'Asian Tigers' have successfully exhibited these abilities. Thus, the inability of indigenous cultures to master or adapt to the language of foreign technology has been one factor responsible for the failure of many development projects that have foisted sophisticated technical equipment or machinery upon the local people. For Salim (1992), the current challenge to all developing countries lies in fostering a synergistic development between science, technology, and cultural values. Developmental programmes in the past have failed because of the fundamental misunderstanding of these cultural peculiarities.

Theoretical Orientation

This study has been partly inspired by cross-cultural theory and associated debates on economic performance over the past three decades. The pioneering works of Geertz, Hofstede, Child, Laurent, and Francis Fukuyama have been particularly influential. For many years, cross-cultural studies have been predicated on psychological experiments and anthropological research (Frijda and Jahoda 1966, Vernom 1965b, Segall *et al.* 1963). In recent times, however, cross-cultural discursive and analytic frames have been developed in hermeneutics, critical theories, paleontology, ethnology, semiotics, structuralism, poststructuralism, economics and management research (Hofstede 1980, Child 1981, Laurent 1983, Smircich 1983c, Foucault 1984, Adler 1991 and Geerts 1993).

In its academic manifestation, cross-cultural theory provides converse propositions to the hitherto simplistic and universalistic assumptions of human nature in modern organisations (Adler, 1991). Organisational behaviour across cultures often transgresses previously conjectured universal limits. It is now generally agreed, for example, that there is no 'global work ethic'. This therefore implies that it is 'unrealistic to advocate or conceptualise universal assumptions of human nature' (Bulcholz, 1977, pp. 567–587). The notion of multiple explanations, or polyvocality, has become a linchpin of cross-cultural theory. In cross-cultural theory, therefore, the emphasis is on particularism rather than universalism in the interpretation of social behaviour. Since organisations often engage in different productive activities, achieve different growth rates, and evolve within different cultural contexts, the application of universal managerial theories could be misleading. From a sociological perspective, cultural variations are crucial in shaping different entrepreneurial and managerial patterns across societies.

For Morgan (1986) historical and cultural variations in societies are reflected in different philosophies and management practices. The environment and its peculiar historical processes influence investment behaviour and industrial relation patterns. The recognition of cultural influence on corporate strategy and economic success has given rise to management terms such as 'Japanese, American, and French management strategies'. Japanese management patterns are typically 'corporatist'. American businesses are oriented towards a free market system, in contrast to state-controlled patterns in France.[18] Chinese, Korean, and Italian businesses are typically family oriented. By contrast, Japanese investment attitudes are strongly geared toward non-kin enterprises (Fukuyama, 1996, pp. 20).

Hayashi (1988) contends that, Japanese values and norms are more compatible with modern enterprises and economic growth than those of their Korean and Taiwanese counterparts. These conclusions were reached through a comparative study of Korean, Taiwanese and Japanese businesses. Whereas managers in Japanese enterprises strictly emphasise the completion of tasks and attainment of objectives, their Korean and Taiwanese counterparts are less strict and more tolerant. The Japanese, he further observes, are more receptive to changes and Western influences, whereas Korean and Taiwanese firms strongly resist external influences. These examples lend credence to the fact that corporate strategies are not the same across the world.

In the cross-cultural analytical approach, a comparison is usually made of the relations between culture and organisation, and the similarities and differences in the attitudes of entrepreneurs and management in different cultures. Cross-cultural research usually attempts to uncover the complexities in group relationships that occur in multicultural work environments. In some cross-cultural studies, the hypothetico-deductive method is applied for generating new hypotheses, while in others, the exploratory method is used for providing new insights into cross-cultural behaviour. In a study of managerial philosophies across nine Western European countries, the United States of America, Indonesia, and Japan, Laurent (1983, pp. 75–96) observes distinct management patterns in each country. He contends that while American managers withhold explanations from subordinates to encourage initiatives, the French, in contrast, consider providing explanations to subordinates as most important. Whereas bypassing the boss is an act of insubordination to most Italians, in the Swedish context, the hierarchy may not be necessary.[19] From his findings, Laurent concludes that culture and national origin influence views on appropriate management strategies.[20]

Hofstede (1980) captures a pervasive aspect of understanding cross-cultural behaviour across organisations. His cross-cultural survey of over one hundred countries, including Nigeria, shows that managers and employees from various countries vary in four dimensions: individualism/collectivism, power/distance, uncertainty/avoidance, and masculinity/femininity role definitions. Subsequent parts of this study examine some of these cross-cultural entrepreneurial and managerial patterns.

Cross-cultural organisational studies are generally open in approach. Thus, from an open system's perspective, 'organisations are influenced and penetrated by their environments' (Scott, 1986, pp. 53). Consequently, the role of actors in the organisation serves as a reflection of societal culture.

John Child's (1981, pp. 303–356) synthesis of organisational studies across cultures provides stimulating results. He observes that the focus on macro or micro level issues has influenced the researchers' conclusions. While studies concluding convergence focused on macro level issues such as the structure of technology, those concluding divergence focused on micro level issues such as the behaviour of participants within the organisation. From this, Child concludes that, while organisations across the world are growing more similar, the people within them are maintaining their cultural uniqueness. The maintenance of cultural distinctness is obvious in multicultural work environments. For international firms operating in other cultures, there is always the challenge of reconciling organisational goals with societal cultural values (Macdonald, 1994, pp. 79–100).

In work situations involving people from different backgrounds, new ways of thinking and behaving often emerge. Cultural diversities in the world create the potential for mutual accomplishment that often supersedes 'the single contribution of each party to the inter-cultural transaction' (Moran and Harris, 1981). The basic tenets of cultural diversity also lead us to the concept of cultural synergy. Fuller (1981) explains that in synergy situations the behaviour of the whole cannot be predicted by the behaviour of any part taken separately, which implies that proper understanding can only be achieved by working from the whole to the particular. For Adler (1991), 'the image of pluralistic, rather than homogeneous, society underlies the synergy approach'. The assumptions of equifinality in synergy explanations imply that there is not just one way of achieving a goal but several ways (ibid.).[21]

To sum up, the cross-cultural approach provides useful understanding on managerial differences across societies and business firms. Organisational behaviour is entrenched within the structure of norms, myths and rituals that make up the organisational culture (Miles, 1980). Organisational culture is a body of customary beliefs and social forms within any given organisation. The different management strategies employed by owners of business enterprises make the study of organisations within a cross-cultural context relevant. Some empirical verifications of this study are drawn from the assumptions, arguments, and propositions of the cross-cultural theory. Nevertheless, cross-cultural studies of organisational behaviours have their limitations. Some limitations of the cross-cultural approach are briefly as follows.

Some Limitations of Cultural Theories

The analytical diversity in cultural study often leads to inconsistent interpretations among researchers. Gerholm (1994) observes that culture is analogous to an iceberg that has its largest part hidden beneath sea level. Culture has become an elusive idea used metaphorically in organisational studies. Metaphorical expressions sometimes score low on the intellectual scale (Alvesson, 1994). Metaphors play double roles. On the one hand, they help to illuminate broader issues, on the other they obscure aspects of particular phenomena by narrowing interest to specific perspectives (ibid.). In the modern organisation, the different cultural backgrounds of participants and cultural misinterpretations of behaviour often account for the internal tensions. Baker's (1980) conception of culture as a 'social glue' could be misleading because cross-cultural misapprehensions often become the source of tension and disintegration in many organisations.

Cross-cultural studies of organisational behaviour often face the complex task of breaking away from macro to micro issues. Comparisons of differences in the behaviour and performance of enterprises are often complicated by the fact that structures, resources, and goals are not the same across business organisations. Although the three firms examined in this study belong to the plastic manufacturing industry, their organisational structures, resources, and manufactured products are varied. Such differences make cross-cultural comparisons of the three firms problematic. Furthermore, in analysing foreign cultures, the researcher's cultural background often sensitises him/her to specific stimuli, which may lead to interpretational biases.

Generally cross-cultural studies of business enterprises are considered to be management-centred. In most cross-cultural organisational studies, the contributions of factory workers and lower groups are relegated to the background.

In order to avoid this common pitfall, entrepreneurial and managerial practices in relation to the labour process are examined. Since entrepreneurs and business managers do not operate in a vacuum, an examination of the way they interact or relate to their employees provides a good understanding of their behaviour. Workers are sometimes useful in providing more objective and better assessments of entrepreneurs and managers under whom they work.

A common criticism against cross-cultural studies is that they lack essential variables that can be measured with certainty (Hofstede, 1980). Often such studies deal to some extent with the cognitive and psychodynamic aspects of human behaviour, such as internalised traits of values and attitudes that cannot be systematically measured quantitatively. A basic limitation of cross-cultural studies is that, they do not often explain the meanings people attach to their behaviour across different cultures.

Cultural and religious explanations often arouse suspicion within academia. Usually, such studies are considered as either ideological or having hidden agendas. The tendency to denigrate cultural explanations as biased therefore constantly looms. Nevertheless, this study uses religious and cultural explanations as analytical tools only because of their role in Kano's current social and economic dynamics.

In recent times, the globalisation and socio-cultural transformations across the world imply that there is no longer any distinct or homogeneous culture. Consequently, 'the discreetness of cultural boundaries has dissolved'. There are now complex obliterations and mixing of cultures and meanings from disparate sources (Friedman, 1995, pp. 5–30) or a 'cultural creolisation' (Hannerz, 1995).[22] The identification of business nationals, such as the Lebanese and Chinese, with specific corporate cultures in this study could therefore be misleading. While corporate cultures exert influence on managerial and employee behaviour, the contributions of the environment and other unidentified variables cannot be ignored.

The use of normative concepts dominates cultural studies. Normative explanations often have empirical limitations. It is not possible, for instance, to measure accurately the salience of a norm in an individual or group unless by observing the individual or group (Hechter, 1987). A basic limitation of cultural explanations is that they point to influences which policies sometimes can hardly reach (Marris, 1968). Human variations across culture imply that what is considered positive in one culture may be entirely negative in another (La-Barre, 1964). Cultural diversities often lead us to different interpretations of a given phenomenon (Morgan, 1986).

An Outline of the Study

This study was first published in December, 1997, as a doctoral dissertation at the Department of Sociology, Uppsala University, Sweden. Since then, much of the information has been updated with three additional Chapters. Expansion of the work with a slight shift of focus necessitated a change of title. The three-part study consists of Twelve Chapters. The first Four Chapters constitute Part One. Chapter One outlines the mainstream understanding, with a theoretical background on culture's relevance to economic performance. A re-examination of various cross-cultural studies can illuminate culturally distinct patterns of entrepreneurship across societies. In particular, Chapter One discusses the method and unit of analyses deployed in the study.

Characterisation of the meanings and conditions necessary for entrepreneurship is central to the discussion in Chapter Two. In the same Chapter, the historical evolution of indigenous entrepreneurship, precapitalist and modern forms of venture capital are reviewed, by attempting to bring analytical sociological perspectives to bear. Entrepreneurial risk perception and its impact on investment strategy in Kano and parts of Nigeria are extensively discussed. Chapter Three builds on the theoretical foundations of Chapters One and Two by attempting to provide a link between culture and entrepreneurial rationality. Chapter Three extends the analysis to include the logic between religious reform and new forms of economic rationalisation in northern Nigeria. Chapter Four focuses on entrepreneurial trust relations. Key conditions and relevance of trust in business transaction are highlighted here. Entrepreneurial network relations and their relevance for the functioning of business activity especially in the plastic sector are examined.

The second part of the study consists of Chapters Five, Six, Seven, and Eight. Chapter Five looks at the research setting. Here, a case by case analysis of the three plastic firms is presented. This is done by first looking at the economic environment within which the plastic firms in northern Nigeria operate. In particular, social networks and clientelistic relations in the Nigerian plastic industry are examined. The impact of recruitment through social or clientelistic networks and its consequences for both bargaining and industrial relations are also examined. The entrepreneurial profit motive is central to the discussion in Chapter Six. Cultural definitions, calculations, and the use of profit receive special attention in the Chapter. The Chapter examines the basic determinants and management of profit in the economic environment. In particular, entrepreneurial profit sharing methods with subordinate factory workers and their impact on trust relations are reviewed.

I begin Chapter Seven by looking at entrepreneurial and managerial resource control methods in the three plastic firms. Differences in entrepreneurial control strategies and the implications for employee and management relations are of particular significance. In Chapter Eight, economic activities in the informal sector are examined. Of particular relevance to this discourse is the role of surplus urban labour in the two sectors of the Nigerian economy. This is done against the background of what is characterised as institutionalised migration. The Chapter reviews, patterns of seasonal labour fluctuation and their impact on the business sector in Kano and northern Nigeria as a whole.

Chapters, Nine, Ten, Eleven, and Twelve constitute the Third Part of the study. In Chapter Nine, the economic role of women and the concept of economic rationality in the Muslim Hausa society are examined. The hidden economic activities of secluded women are central to the discussion in this Chapter. The framework of analysis in Chapter Ten is extended to cover the subject of business success. In addition, entrepreneurial cross-cultural definitions of business success and their impact on investment patterns are examined. The various managerial strategies for motivating workers and their impact on the smooth operation and overall performance of enterprises are also examined. Chapter Eleven looks at economic development by reviewing the various dependency and modernisation paradigms. In this Chapter, an attempt is made to reformulate the modernisation theory from cultural vantage points. The Chapter among other things looks at corruption and business fraud and their implications for economic development, in Nigeria. The major implications for trust in the business sector and the nation as a whole are extensively discussed.

Chapter Twelve, which is the concluding section of the study, provides a summary of empirical findings. The study is concluded through a general assessment of entrepreneurial ethics and trust related attitudes in the northern Nigerian business sector. The final section of the study examines the contribution of entrepreneurs to economic progress in northern Nigeria and offers suggestions on ways of addressing contemporary problems particularly as they relate to trust and economic performance.

2 The Making of Indigenous Entrepreneurs

> My ventures are not trusted to one bottom, nor to one place; nor my whole estate upon the fortune of this present year: therefore my merchandise makes me not sad (Shakespeare *in, The Merchant of Venice*, Act 1, Scene 1).

This Chapter looks at the historical evolution of entrepreneurship in northern Nigeria. Much of the analysis regarding the evolution of venture capital, entrepreneurial ethics, and risk-related attitudes has relevance to the concept of culture discussed in Chapter One.

Research on the role of the entrepreneur in society has become a highly controversial exercise. This is because entrepreneurship is a complex labyrinth which involves different frameworks of activities (Haahti, 1989). In spite of the diverse approaches often applied in the study of the entrepreneur, there is no generally accepted method for capturing the entrepreneurial behaviour. For economists of the Austrian tradition, the term entrepreneur refers to actors in the marketplace (Woodward, *et al.* 1994, p. 3). The entrepreneur in this context is regarded as an optimiser, an individual who combines resources or factors of production in an optimal fashion (Kirzner, 1960). In the works of Hayek (1980a, [1945]) on the other hand, the entrepreneur is described as a prime innovator and an instrument of change. Essentially, entrepreneurship is about change. Conversely, obstinacy is antithetical to the requirements of entrepreneurship. The entrepreneur identifies possible economic opportunities and outcomes in the same environment in which others may not.[23]

Ostensibly, research findings on the role of the innovative entrepreneur are, at best, parallel. Whereas, for Kirzner (1983, pp. 281–291), innovative entrepreneurs create a new equilibrium through alertness and the ability to grab unutilized opportunities, in the works of Schumpeter (1962) they generate a disequilibrium through 'creative destruction'. There is no predictive entrepreneurial trait. However, what seems to cut across the board among a broad spectrum of successful entrepreneurs is the willingness to undertake risk and to persevere. The essential links in entrepreneurial analyses are risk and perseverance. The entrepreneur exists only because of 'uninsurable risks'[24] and human willingness to persevere. Risk-assuming and resilient entrepreneurs are therefore found in both the public and private sectors of the society. This study, however, places emphases on the activities of private sector entrepreneurs.

Unlike public sector entrepreneurs who are key policy makers, private sector entrepreneurs, the main agents of economic transformation, have the market as their arena of operation. Private sector entrepreneurs play leading roles as inventors of new ideas, products, and services in the society.[25] In view of this, entrepreneurs are both business people and owners of enterprises. The analysis deployed in this text departs from the idea of the transient role of the entrepreneur.[26] Conversely, the state of entrepreneurs is seen as dynamic, reinforced by constant struggles and adjustments to keep businesses on track. Although modern entrepreneurs sometimes share managerial functions, they are considered as distinct from the managers.[27] In the modern organisation, managers often help entrepreneurs in the execution of delegated functions. Management, therefore, is the art of getting things done through people (Mescom, *et al.* 1985).

Most of the transformations and changes that occur in the economic life of societies are largely due to the initiatives of key individuals or groups from within. Societies that lack local initiatives often stagnate. Schumpeter (1962) observes that the growth of an economy is not natural or inevitable, or even steady, but rather pushed forward in sudden leaps by the activities of key actors in society. The individuals or groups who undertake such economic risks are usually people with high 'achievement motives' whose quest to succeed knows no bounds.[28] Generally, people in pursuit of distinction often appear eccentric in their own societies.

Patrimonial or neo-patrimonial domination and access to the state are key factors in entrepreneurial success in many developing societies, including northern Nigeria. The evidence, however, suggests that the economic prosperity of the entrepreneurs in Kano preceded the prosperity of the modern Nigerian nation and the oil boom era. Successful merchant capitalists, some of whom later diversified to industrial activities, were 'self-nurtured' people. Post-independence state support neither created indigenous entrepreneurs, nor provided grounds for their emergence, rather it provided the clientelistic network which helped to consolidate the position of aggressive merchants in the modern political economy of Kano. In most of the illicit economic alliances between state bureaucrats and indigenous entrepreneurs, the former merely exploited the latter to siphon off funds from state contracts.

In Hausa society, material wealth is closely linked to social status. In view of this, adept and successful entrepreneurs are admired and respected. Material wealth consists not only of accumulated wealth and property, but also of one's level of influence or control over other, less fortunate members of the community and primordial groups.[29] Richness, in terms of control of men, is very important in enhancing the prestige of a successful person (ibid.). An entrepreneur's ability to control extensive social capital goes a long way in boosting the size of his/her clientele, and outside support in times of economic misfortune. In their bid to gain control over less fortunate members of the society, local entrepreneurs spend a sizeable amount of their profits solving social problems around them. Successful Hausa entrepreneurs who fall short of these expectations are often regarded as social deviants.

Although entrepreneurial generosity may be appropriate in a non-economic social context, it is sometimes unhealthy for capital development. Garlick (1971, p. 146) for instance, considers African social structures and the burden of supporting the extended family and primordial groups as important obstacles to entrepreneurial advancement. Kinship obligations are the conduit pipes through which the resources of an incipient businessman are siphoned.[30] This however, does not imply that indigenous business operators do not manage to keep family and kinship obligations within reasonable limits. The evidence in Kano suggests that they do.

The labour-intensive character of mercantile activity explains the relevance of control of many men by a single entrepreneur. Usually, business activity is conducted through chains of dependants and networks of trusted clients within the same city or across the country. In the business sphere, dependants, otherwise known as *yara* or *barori,* are sometimes very useful in providing their merchant patrons with relevant information on market trends. Furthermore, they often perform useful roles as helping hands. Such persons sometimes do not have fixed wages, but depend on occasional support or *quid pro quo* from successful entrepreneurs for earning their livelihood.

Religious tenets encourage a modest display of wealth because of the belief that divine bounty and gratitude go hand in hand. Society expects affines, consanguines, and neighbours of the affluent entrepreneur to benefit from his economic success. Because of these expectations, a wealthy person finds it difficult to refuse to lend financial support or to guarantee a bank loan (Grégoire, 1992, p. 146). In spite of these fundamental expectations, Islam condemns both excessiveness and miserliness. The Qur-an portrays spendthrifts as 'companions of Satan'.[31] On the other hand, it condemns miserliness of wealth as a 'scandalous act', especially where the needy are in overwhelming numbers. This is because miserliness blocks up charity, or *sadakat* and the 'circulation of goodwill among men'.[32] The position of the Qur-an on the behavioural conduct of the affluent Muslims is a moderate one. It encourages a measured degree of generosity, without excessive miserliness or prodigality (Rodison, 1980).

Asceticism is not a distinct feature of a specific environment but a universal characteristic among business groups across the world. In Kano and parts of northern Nigeria, entrepreneurial temperance, thrift, and ability to undergo initial processes of self-denial is a prerequisite for business success. This self-denying process often marks the beginning of capital acquisition. Adam Smith (1981 [1776, 1789]), has observed that entrepreneurial parsimony and frugality were the starting points of capital formation and accumulation, whereas extravagance and prodigality, he observed, dissipated and retarded the growth of capital.[33] I agree, along with Tahir (1975) that there is a direct rather than an indirect link between cultural and religious factors and economic behaviour among the Muslim entrepreneurs of Kano. Many, if not most of the indigenous Hausa entrepreneurs, perceive economic activity as an extension of religious practice.

The resurgence of indigenous entrepreneurs in Hausaland, and especially Kano, can be categorised under three main historical epochs. These are the pre-colonial, colonial, and post-colonial eras. The combinations of long-distance commerce and a vibrant local trade have been important ingredients of the city's relatively prosperous economy for at least four hundred years before 1912, a period that marked the integration of Kano into the European world economy (Hogendorn, 1978). Before this integration, Kano served as the chief *entrepôt* to the trans-Saharan trade.[34] Long distance trade links or *fatauci* with the Southern forest areas extended as far as Ashanti and Gonja[35] in the former Gold Coast. The main commodities of trade include grains, desert salts, natron, indigo-dyed textiles, cola-nuts, and cattle. Although commercial activities over the vast expanse of the desert and with the forest hinterlands were difficult, they were ideal for the initial development of merchant capital.[36]

The introduction of British colonial rule in Nigeria in 1903 marked a change in trade patterns. The creation of national boundaries led to the severing of old commercial linkages. Conversely, new trade routes were created through the Southern coastlines of Nigeria to provide contact with Europeans. The opening of the Kano railway, in 1912, was the beginning of the link between the social formations of northern Nigeria and the capitalist world. This economic relation has remained unbroken till today.[37] The construction of new roads and rail networks during the colonial era further enhanced economic activities between the northern and the southern parts of the country. Colonialism introduced a more organised monetary economy. Hitherto, silver, gold, and cowry currency circulated side by side. Colonial encouragement of the local production of export crops altered subsistence agricultural patterns throughout Hausaland.[38]

The boost in groundnut exports between 1906 and 1914 was a mixed blessing that brought both wealth and poverty to Hausaland. To the relatively small merchant class and state bureaucrats, cash crop exports meant economic prosperity, whereas, for most peasants, or *talakawa*, the heavy reliance on cash crops for export implied poverty and decline in the fundamental local food production of grains and tubers. Frequent depletion of local food reserves during the cash crop era implied that basic foods often had to be sought elsewhere.[39] In any case, the real beneficiaries of the groundnut boom were mostly European buyers, Levantine and Syrian communities, a small class of indigenous merchants, and their protégés. The merchants who were intermediaries in this trade merely exploited price differentials between foreign buyers and the local peasant farmers through a complex network of clientele. The European buyers depended on Levantine and indigenous middlemen for the purchase of cash crops such as groundnut and cotton from rural farmers.[40] The intermediary role of the Lebanese in business activities made them the closest competitors to 'the immediate aspirations of the local merchant class' (Olukoshi, 1985, p. 61).

The era of cash crop exports, with its symbolic groundnut pyramids, was significant in the sense that it was a historical epoch that saw the emergence of new entrepreneurial groups in Kano. A relevant feature of the commodity trade was that it brought rapid economic gains to the local merchant capitalists. Watts (1987) observes that the colonial policy of indirect rule in northern Nigeria 'and its trade-based political economy greatly expanded the dominance of a local prebendal ruling class and opened avenues of accumulation for Hausa merchant capitalists'. The commodity trade therefore is coterminous with the consolidation of Kano's position as the leading commercial centre in northern Nigeria. By the early 1960s, the groundnut trade had brought unprecedented wealth to the city of Kano hitherto unknown in many parts of Hausaland in the West African subregion. In the 1960s, the prosperity of Kano led to a burgeoning demand for European goods such as clothes, cars, household electronics, and imported food items. The trend led to the expansion of the commercial and service sectors of the economy.

The cash crop economy and the towering groundnut pyramids of Kano were on the verge of disappearance by the early 1970s, due to the stagnation of producer commodity prices and the new boom in the oil sector. The large amount of revenue generated from oil during this period led to new government-funded development projects in many parts of the country. In Kano, significant changes during the oil era include the creation of industrial layouts, the construction of new markets, expansion of road networks, health, education and capital-intensive irrigation projects.[41] The oil boom encouraged a tremendous inflow of foreign capital investment in major Nigerian cities such as Kano, Lagos, Port Hacourt, and Kaduna. The major foreign capital investors in Kano during this period include Belgian Eternit, Dunlop Foam, Societé Commerciale Ouest Africaine (SCOA), textiles, Fiat Trucks, Raleigh Bicycles, Imperial Chemicals, Union Carbide Batteries, Leventis Bottling and Motors, Lonrho Milling and Tanning, Texaco, Total and Shell. There were also appreciable foreign investments in the telecommunication and banking sectors with the Bank of America, Chase Merchants Bank, and the Bank of Credit and Commerce (BCC) leading the trail of foreign financial institutions.[42]

With the new boom from the oil sector, many peasant farmers of agricultural export crops abandoned arable fields in search of better-paying factory jobs in the cities. This new trend saw a large-scale labour migration from the rural to urban centres, with an abysmal rate of inflation resulting in a decline of the agricultural sector. Consequently, the peripheral position Nigeria occupied on the world economy suffered further decline. With the new economic changes, the merchants of Kano, intermediaries in the cash crop trade, diverted their energies into building and road construction projects sponsored by government at the federal and state levels. The indigenisation policies of the Federal Government of Nigeria during this period enticed local entrepreneurs into buying shares in many expatriate firms and diversifying into small and medium-scale industrial activities. In recent years, however, the poor performance of the oil sector, coupled with failure of many indigenous manufacturing and processing firms, have led to considerable reversions to commerce.

The history of entrepreneurship in Kano, like elsewhere in Africa, is dominated by opportunism and swift mercantile reactions to prevailing lucrative economic activities. As such, there is often a reluctance to break away from 'historically generated predicaments' (Himmelstrand, *et al.* 1994, pp. 16–36). There is an absence of a genuine interest in boosting local food production or integrating the primary sector with capital goods. Shenton (1986) observes that, even if northern Nigeria was integrated into the capitalist world, it did not develop the forces of production or develop Western capitalism. By its parasitic nature, merchant capital in Hausaland created acute shortages because it exploited and marginalised peasant farmers while it failed to encourage real food production.[43] The universal capitalist approach, with its symbolic mass production, has certainly been lacking. Consequently, the germination of 'petty capitalists' has been restricted to commerce (Hopkins, 1973).

Although merchant capital existed in the pre-colonial and post-colonial era, its ability to transform the commercial sector into a more productive modern manufacturing sector has remained a big challenge. Nevertheless, the most common argument for the low participation in manufacturing ventures hinges on the inadequacy of capital (Akeredolu-Ale, 1977). Many Nigerian entrepreneurs are seen as lacking the requisite capital capable of launching a meaningful modern business enterprise, especially in the manufacturing sector. Nevertheless the hypothesis of capital shortage is unsupported, according to Schatz (1977, p. 67) and others. Perceptions of lack of capital arises 'when an enterprise or an aspiring entrepreneur seeks capital for a venture that does not have a reasonably good chance of business success' (ibid.). Although the inadequacy of venture capital might still be one of the major constraints, this is sometimes contradicted by the non-participation of highly successful merchants in the industrial sector.

Venture Capital and New Businesses

In many areas of capital analyses there are discrepancies and often ambiguities about theoretical and empirical measurement (Lachmann, 1956, and Ward 1976). Measurement problems also arise because 'the same term is used to serve different purposes'.[44] Whereas, in classical economics, capital is separated from other factors of production, such as land and labour, in the neoclassical economic context, there is no separation between capital and other factors of production. In the daily economic analysis, capital is used interchangeably with material resources of production and sometimes their money value. Capital is thus understood as 'the present value of income and future income streams'. The difference between capital and income is that while the former is stock, the latter is associated with a notion of flow.[45] Inherent in capital analysis is the '*ex post*' and '*ex ante*' notions of durability and potential for yielding future income benefits (Ward, 1976, p. 19). In Marxists analysis, this 'self-expanding value' of capital is expressed as 'money generating money'.[46] Capital, therefore, has a relevance to labour inputs in the production process.

It is, however, significant to mention that in Hausa society, capital, or *jari*, which has several uses, generally denotes funds available to a businessman for any form of economic transaction. This form of capital is not fixed, but is convertible and sometimes transmitted to succeeding generations.[47] Like many dual economies, precapitalist or traditional forms of venture capital not only coexist, but compete side by side with modern capitalist banking systems.

Research on sources of capital for financing new business shows that personal savings are most important (Thorne and Ball, 1981; Vesper, 1980). Next in order of importance is financial support from family and friends (Litvac and Maule, 1973). The results of the study, however, show a nearly opposite picture. Apart from indigenous Hausa entrepreneurs, among whom 71 percent mentioned personal savings as an initial source of capital, the corresponding respective figures of 20 and 33 percent of Lebanese and Chinese entrepreneurs mentioned personal savings as a source of capital. Significantly, 60 and 70 percent of the Chinese and Lebanese business proprietors respectively got their initial capital from family sources. Loans from banks scored lowest among all business groups, representing less than one-fifth of entrepreneurial capital. Most Hausa and Lebanese entrepreneurs in the empirical survey, 61 and 80 percent respectively, defined capital as 'initial money used to start a business activity'. There was no significant difference among Hausa, non-Hausa and Lebanese definitions of capital, whereas 60 percent of their Chinese counterparts perceived capital as 'human and material resources used in the running of a business'. The initial capital, or *jari*, for a venture is usually insufficient to make a meaningful impact. Therefore, successful businesses are usually the products of several attempts and failures (Day, 1986).

In line with the two major sources of venture capital identified among the business groups of Kano, a Hausa business person in search of commercial capital resorts to personal saving methods known as the *adashe* or *asusu*. The *adashe* is a credit-rotating scheme in which a group of traders or people in search of capital forms a cooperative union by contributing an equal amount of money in a pool. Gathering money from different persons allows each individual a rotating credit facility. This process is repeated at intervals until each person in the group has had his turn (Bashir, 1986, p. 2). A major advantage of this method is that the contributions made by many traders provide an individual the opportunity to have enough capital to be capable of starting a new business, a process which would have been impossible with purely personal savings. The credit-rotating scheme also has the advantage of fostering solidarity among business groups.

The *asusu*, which translates as 'money-box', is a traditional saving scheme in which an individual undergoes self-denial to form or accumulate enough capital for a business venture. The process, which is repeated daily, sometimes involves keeping money in inaccessible places or giving them to trusted persons for safe keeping until the money has accumulated enough to serve the purpose for which it is intended. The *gayya,* a form of covee labour, is also traditional capital generating device. This method is largely practised among peasants in the rural areas. In recent years, the method has lost popularity, especially in the cities. In the *gayya* method, a group of persons or families supply free labour to an individual on the farm (Bashir, 1986). The advantage in this method is the supply of free labour, which saves the cost of hiring farm labour, or *Yan kwadago*.

In the traditional method, there are often no fixed or formal rules for credit provisions among the local business groups. In the absence formal rules, credit allocations are often not legally binding. This implies that in spite of the usefulness of the traditional micro-credit schemes, they are beset with operational problems. Sometimes uninsured money-boxes containing a person's life savings can be stolen or lost in natural disasters, rendering individuals bankrupt.[48] Although a person in search of capital loans from a successful merchant could mortgage a piece of land, animal, or building. Such pledges or local mortgage procedures are known as *jingina*. On the other hand, land as a form of capital can be acquired freely from local functionaries, lent *aro* or sold on a nominal fee.

Unlike commercial capital, which is often small, extensive or large capital necessary for the launching manufacturing activity often comes from long-term savings or in form of loans from the modern banking sector. Other sources from which capital is generated include company shareholding, working as contractors in government-funded projects, dealing in petroleum products, or acting as agents to multinational firms. Most indigenous business people in Kano serve as brokers, or *dillalai*, to modern business firms as a means of getting capital. Company shareholding became fashionable among indigenous entrepreneurs in the 1970s with the introduction of the federal government's indigenisation policies. The policies were introduced as an attempt to encourage indigenous people to participate in and control economic activities.[49] Despite these initiatives, Nigerian governments have not successfully transferred substantial capital into the hands of local entrepreneurs. Beckman (1982) explains that 'the suggestion that the Nigerian state promotes capitalist development does not necessarily mean that it does so efficiently. Nor does it mean that such development provides useful answers to the problems of mass poverty and deprivation facing the Nigerian people'.[50]

In recent years, rentier capital has become a lucrative source of wealth accumulation among indigenous entrepreneurs. With the new boom in the mortgage business, the building and construction sector offer greater economic opportunities. Capital for many investors is acquired through rents on housing property. Entrepreneurial borrowing from banking institutions and lending agencies became more noticeable in the early 1950s among a new generation of the indigenous business community who were actively engaged in the commodity trade. Nevertheless, the banking institutions are often unpopular with the older generation of local merchants due to Islamic views concerning interest rates on loans. Despite this, the loans from banking institutions are the biggest source of capital for modern business ventures. It is significant to mention that, although very modest and sometimes cumbersome to operate, the traditional methods of generating capital are still widely practised among the commercial groups in Kano.

Merchant capital is based on the exchange of goods and services, a trend that has become a pervasive method of capital formation among the entrepreneurial groups in Kano. Profit is realised through the appropriation of the difference in the value of commodities and from unequal exchanges (Shenton, 1986). The innovative acumen of Kano traders during these years has been restricted to their abilities to exploit and maximise price differentials or discrepancies between the buying and selling points. Traders sometimes hold goods, especially grain and leguminous crops, until the market has fluctuated from low to higher prices. To be able reap the advantages of regional price differentials, the trader or merchant must be alert to changes in the market forces. Most Kano businessmen display the quality of 'alertness' in abundance for exploiting price differentials. Kirzner (1973) recognises 'alertness' as an essential quality of the entrepreneur.[51] However, 'alertness' alone is not a sufficient condition for entrepreneurship. The assiduous entrepreneur also requires a considerable amount of judgement on several impinging variables, such as risk, potential demand, and the reliability of supply lines (High, 1982).

Mercantile or trade gains, as Belasco (1980) has noted, were not derived from any changes in production relationships. Trade gains do not depend on surplus labour and product costs, but on selling price increases. Indigenous capital has therefore been restricted in the monetary sense and not in capital goods. Illife (1983) explains that Nigeria typifies the West African model of a nation growing out of pre-colonial exchange that also survived the impact of colonialism. Capital among business groups was first rooted in trade and later diverted to agriculture and industry.

Levine (1966) is particularly of the view that participation of Hausa merchants in indigenous capitalist expansion is unsatisfactory compared to their Ibo and Yoruba counterparts. In ranking the achievement motives of the three major ethnic groups in Nigeria, Levine placed the Hausa achievement motive as the lowest. The reason, according to him, is that, unlike the other ethnic groups, Hausa society is hierarchical, with upward mobility based on 'loyalty, obedience and sycophancy' rather than individual achievement. The evidence in Kano and parts of Nigeria however, suggest that Levine's study could be misleading because it failed to properly address the different cultural values among the ethnic groups.[52] A common tendency to denigrate the initiative of Hausa people may be fallacious judging from the historical economic activities of the group and their past and current roles in Nigeria's economic life (Hogendorn, 1978). Throughout sub-Saharan Africa, the impact of the economic contributions of Hausa emigrants is noticeable in major cities such as

Abidjan, Accra, Kumasi, Freetown, Banjul, and Conakry. Akeredolu-Ale (1975) debunks the notion that entrepreneurs with formal higher education are more successful in business than their less-educated counterparts. Pickles and Farrel (1986) caution that the relationship between entrepreneurial achievement and education is highly complex.

The lower education of entrepreneurs compared to other groups has been observed to be a common trend in many societies (Kent, *et al.* 1982). Generally, people with entrepreneurial motives are often less favourably disposed to formal education, which requires large investments in time. It is true that some of the ablest and most prosperous entrepreneurs or businesspeople in Kano and other parts of Nigeria have never been to school.[53] Sometimes uneducated entrepreneurs in Hausa society have succeeded on subsistence incomes that their educated colleagues would have considered inadequate for business transactions. A study by Brockhaus and Nord (1979) reveals that the level of education of entrepreneurs was generally lower than their managerial counterparts. Marris (1968) contends that some of the most successful businesspeople in Africa are those whose bookkeeping methods are often haphazard, with speculative costing, and notes jumbled on paper scattered across their offices. This implies that, although education may enhance efficiency and the ability to operate large-scale industrial manufacturing activities, it does not necessarily determine success in petty business activities characteristic of the competitive informal sector of Kano and sub-Saharan Africa.

The Colonial Epoch and Indigenous Entrepreneurs

Historical records and radiocarbon dating from archaeological artifacts show that a variety of mercantile, craft and local forms of manufacturing activities thrived in Kano and its neighbouring districts before contact with the colonial world in the latter 19th century. Traditional industries rooted in local skills are useful in allowing society to develop technology from cultural vantage points. Early traditional industries include cloth weaving and dyeing, tanning and leather works, and the fabrication of metals for domestic household use, agricultural tools, and weapons. The handloom was particularly well developed before the landmark encounter with the outside world. Kano and its rural districts provided the cloth needs of nearly half the central Sudan in the 1890s. The popularity of the 'Kano cloth' was reflected by its ubiquitous nature in sub-Saharan Africa. Indigo-dyed clothes and tanned leather formed a major export to Brazil and North Africa during the period under review. The city of Kano became a predominant textile centre for West Africa and further afield. The city lost this role with the inception of British colonial rule in 1903.[54]

With the inception of colonialism, urban centres of production including the city of Kano in northern Nigeria, became collecting centres for export crops. Colonial introduction of forced-labour led to changes in social relations of production. Many peasants from the countryside were often forcefully engaged in the construction of railways, roads, and prisons. These developments further encouraged internal migration and the displacement of local populations. Colonial creation of national boundaries and the restrictions on the flow of human traffic meant that some important business contacts and capital, as monetary advances which characterised the payment of goods and services in the long distance trade, were lost.

Colonialism also introduced modern technical equipment and affordable, mass-produced industrial goods. Calico and secondhand clothes replaced locally made hand-woven clothes, which were costlier and much slower to make. Moreover, colonial regulation and restriction of the importation of raw materials, such as the dyes needed for the manufacture of local textiles, weakened indigenous initiative in this field (Mustapha, 1985, p. 34). The imposition of the cultivation of certain types of cotton, which was favoured by foreign firms, inhibited to the cultivation of the indigenous cotton and the expansion of the textile industry (ibid.). With the inception of colonial rule, refined factory salts replaced indigenous domestic salts. Local soaps made from palm oil and wood ash lost their importance as detergents with the introduction of industrially manufactured soaps from overseas. The local capacity to adapt or master the language of modern technology was hampered by the rather low levels of literacy among the indigenous artisans and craftsmen. Given this development, local competitiveness and resistance to the encroachment of foreign goods into local markets were weak, or often, absent.

Productive economic activities were sometimes completely abandoned because they were branded as 'inferior', 'crude' or 'primitive' by both foreigners and their local imitators. The preference and insatiable taste for foreign goods weakened the main fabric of indigenous innovation. The colonial Land Proclamation Act of 1910 eventually halted indigenous investment and subsequent development of private land property.[55] Consequently, the local impetus for indigenous industrial takeoff as experienced in Europe and the United States of America was shattered. The prominent role those indigenous merchants played in the local and regional market was replaced by an unequal economic partnership with the capitalist world. In a very erudite analysis of the 'historical causes of Africa's economic underdevelopment', Himmelstrand observes that:

> . . . the interruption of otherwise natural improvements of indigenous entrepreneurship and mercantile capabilities as a result of the fact that such entrepreneur-

> ship and mercantile endeavours were virtually monopolised by the colonialists. At political independence, therefore, virtually no indigenous, national bourgeoisie was in existence in these countries (Himmelstrand, 1994, p. 23).

In the post-independence era, the merchant and entrepreneurial classes in Kano continued to perform their newfound roles as buying and selling agents of cash crops. Invariably, the local merchants were reduced to the role of parasites in economic relationships of which they had little control. Writing about the role of state and capitalist development in Nigeria, Beckman (1982) cogently argues that:

> The colonial state systematically obstructed the growth of a powerful domestic bourgeoisie. It sanctioned the monopolistic practices of the colonial commercial firms in their struggle to eliminate 'native' competition in the all-important import-export business. Politically, it allied itself with backward aristocratic elements blocking the demand for bourgeois forms of political representation (Beckman, 1982, p. 42).

The position of the local entrepreneur has not changed although there has been improvement with the oil economy. This is not to suggest that, without colonialism, local entrepreneurs would have developed, or attained levels comparable to their counterparts in the industrially advanced world. Such an argument would be most difficult to defend empirically. However, what is being stressed here is that colonial intervention has blocked the natural course of indigenous entrepreneurship in northern Nigeria and elsewhere in Africa. The negative effects of the intervention of foreign technology becomes obvious, when analysed within the context of the historical development of the indigenous craft industry.

Craft activity in Hausaland has been diversified since time immemorial. Activities in this field include weaving, leather tanning, pottery, blacksmithing, and woodcarving. Two categories of craftsmen have been identified in Kano and other parts of northern Nigeria. These consist of seasonal and professional craftsmen. The seasonal craftsmen combined craftwork with farming activities. Such craftsmen were only active during the dry and idle season. The professional craftsmen, on the other hand, are people permanently engaged in craftwork as a full-time occupation. The contributions of blacksmiths has also declined considerably from their predominant-colonial importance (Jaggar, 1973).

Historically, the early indigenous settlers of the Dala area of Kano were blacksmiths. A very important component of metal fabrication was the mining of iron ore from local laterites and magnitites. With the inception of colonial rule, indigenous mining activities became illegal. The importation of cheap iron components from abroad further weakened the local mining of iron ores. Colonial prohibition of local mining activities had a negative impact on indigenous technology and the fabrication of tools. Consequently, many craft producers were reduced to mere distributors and peddlers of imported commodities that they formerly produced.[56] Today, the very marginality of craft occupations discourages heirs from taking up the modest enterprises of their parents (Waldinger, *et al.* 1990). Although operating on a smaller scale, the local blacksmiths and jewelers continue to produce indigenous ironware such as doors, beds, locks, hinges, farm implements, bracelets, necklaces, and household tools.

Ironically, national independence has done little to restore the autonomy of craft operators or to protect them from the current economic challenges. The craft industries did not have the resemblance of proto-industries in Europe, America, or Asia. A major handicap of the craft activities in Nigeria has been their inability to bind together as cottage industry (Sokolski, 1965). Individual ownership and control therefore dominate craft occupations. Family monopoly and the hereditary nature of the craft professions in Hausa society limit their ability to grow into modern occupations. Consequently, craft skills are often reduced to traditional titles or are jealously guarded as family secrets.[57] New entrants into craft occupations without previous social links are considered as 'intruders' and are strongly resisted by professionals. Usually, the intruders into craft professions are accused of professional malpractice and the production of substandard goods.

Today, occupational barriers in craft-related activities are getting weaker due to the introduction of advanced foreign technology and the emergence of modern artisans and craft professionals such as mechanics, electricians, welders, and photographers (Grégoire, 1994, p. 34). The indigenous craft industries have been supporting enterprises from family savings. Their concealed nature prevented non-family or outside support.[58] Shea (1974) explains that attempts by the Kano merchant capitalists during the colonial era to infiltrate the activities of cloth dyers in a nearby Kura village were met by strong resistance from the local producers.[59] The inability of the craft industry to mature beyond subsistence levels and its failure to develop into a modern industry is due to its resistance to outside labour and financial support from non-family members. Consequently, cloth manufacturing could not go beyond the technology of the handloom. Most of the craftsmen rarely possessed the skills or desire to create anything but a small firm (Rowe, 1971).

The lack of efficient marketing strategies or networks for products also limits the expansion of craft activities. Craft occupations also rely on a combination of local and foreign raw materials, including cotton and imported chemical dyes for looms; animal skins and hides for the leather industry; wood for the local furniture industry; and scrap iron for blacksmiths. Seasonal and climatic changes such as droughts and famine, combined with government restrictions on foreign imports, frequently disrupt the flow of raw materials. This scenario often keeps the operators of local craft industries out of business for most of the year. Faced with external and internal competition from modern manufacturing firms, the market share of the craft industry appears to shrink daily. Although some of the local craft industries have undergone changes and qualify to be seen as modern, their management and role in the community are not much different from those of the old craft activities.

The inability of the local craft industries to draw support and capital from the Kano merchants was largely responsible for their decline. Second, the intervention of advanced foreign technology further weakened the ability of local craft industries to develop into large-scale modern economic ventures. In Kano, as in other parts of Nigeria, the stagnation of craft and other business activities has been fostered by entrepreneurial attitudes towards risk. A subsequent part of this text examines entrepreneurial risk predisposition and its impact upon investment patterns.

Entrepreneurs and Risk Perception

Entrepreneurial risk perception refers to the process by which investors screen, select, organise, or interpret stimuli or economic signals, so that these stimuli or signals have particular meanings for them. However, because of the many stimuli or economic signals present in an environment at any given time, entrepreneurs tend to screen out most of what they see. Consequently, personal interests, values, and culture act as filters which can lead to the distortion, blocking, and the creation, of what actors in any system choose to see or hear (Adler, 1991, pp. 63–91). Stimuli selection and risk perception depend on many antecedent factors, which include values, norms, goals, personal interests, etc,. (Schoemaker, 1993; Ritchie, and Marshall, 1993, p. 35). Cultural values exercise a significant influence on entrepreneurs' risk perceptions and their subjective attitudes regarding investment decisions.[60]

Risk perceptions are learned and they usually develop over time. Cognition, on the other hand, refers to a set of categories that guide perception and thinking. Individuals use cognition to perceive, classify, and interpret events (Sackmann, 1991). Cognition is therefore an aspect of 'organised knowledge'. Organised knowledge provides the basic frame of reference and guide to which human beings conduct their day-to-day affairs.[61] In many African countries, including Nigeria, the entanglement between political risk signals and cultural variables exerts a significant influence on entrepreneurial trust, risk perceptions, and investment attitudes.

Differing connotations are often associated with the idea of risk across cultures, business groups, and academic discourses. Thus, risk perception from an ecological perspective differs greatly from an economic interpretation. To the business entrepreneur or manager, risk is associated with capital investment, business decision, and technological and social innovation. For the industrial worker, risk may take the form of the probability of losing a job or exposure to work hazards and injury. Since risk has different meanings for different groups, responses often received from risk-related questions are greatly varied. This reality was apparent in the plastic factories of Kano. The responses received from entrepreneurs and management of firms differed entirely from those from the workers they employ. While approximately 43 percent of the entrepreneurs were preoccupied with the 'uncertainties in capital investments', 46 percent of their employees interpreted or perceived risk in relation to 'job insecurity'. Thus, trust or faith shapes risk behaviour in a particular system.

Risk, which in this context implies the extent of willingness to experiment, varies among individuals and groups. Thus 'risk is measured by the probabilities that will reflect changes in the occurrence of possible outcomes' (Thornhill, 1989).[62] Risk as a concept is a theoretical abstraction that relates to specific situations, stimuli, or occurrences considered as dangerous (Ritchie and Marshall, 1993, pp. 25–26). In dealing with risk perception, previous experiences are very important in shaping distinct realms in cognition and behaviour. Risk perceptions are dynamic, often varying across age, gender, and education (ibid.). Yet, risk differs from uncertainty. The difference between risk and uncertainty is that, in the latter, lack of data renders probability calculus impossible. Consequently, intuition and personal judgement pervade the terrain of human uncertainty. Therefore, individual praxis is very essential in risk decisions. Entrepreneurial risks entail a variety of financial, career, family, and psychological sacrifices.

Financial risk is associated with uncertainty relating to the future prospects of personal capital and monies borrowed from other sources to start businesses. Family risk entails sacrificing family interests and pleasures to embark on long distance journeys, which sometimes involved prolonged absence from spouses and children. Psychological risk, on the other hand, refers to the traumas a businessman or woman has to undergo in the face of uncertainties and speculations in the business environment. Personality and biochemical theories maintain that human behavioural patterns towards risk are innate or ingrained in early life; therefore, some people are more naturally prone to take risks than are others.[63]

In cultural analysis, the individual does not operate in a vacuum, which implies that risk perception is influenced by the values and norms of the larger society. The risk-as-value theory argues that there is a cultural value attached to risk taking.[64] Cultural values sensitise entrepreneurs to different opportunities and to dangerous economic signals. Conversely, a waste product in one business sphere could be a raw material in another. Culture produces within its realm both early and late identifiers of opportunities and dangers. Mary Douglas (1985) observes that, in dealing with risk-prone behaviour, it is better to focus on social influence than attempt to eliminate it.

Hofstede's (1980) cross-cultural managerial study of over 100 business firms across the world reveals variations among cultures in uncertainty avoidance. Entrepreneurs and managers in some cultures, he observes, tolerate risk more than others (ibid.). In some cultures, investors try to avoid risk once modest fortunes have been achieved (Sulivan, 1987). Entrepreneurial predisposition towards risk depends on the willingness to experiment in unexplored avenues. The amount of time spent in arriving at decisions also varies among groups. While in some cultures decision making is a quick process, in others it may be slow or completely avoided. In analysing group decision-making in Japanese management, Takamiya (1972, pp. 183–196) and Fukuyama (1996) contend that, because of the several hierarchies in their decision processes, Japanese mangers are slower in reaching decisions than their Western counterparts.

The Hausa conception of risk is similar to the classical economics treatment of risk, which became dominant in the seventeenth and eighteenth centuries. In the classical economic context, risk is synonymous with gambling. This theorising about risk in economics changed in the nineteenth century to a new idea of economic endeavour involving profit and loss (Douglas, 1985). In Hausa language, risk, or *kasada*, is used interchangeably with gambling, or *caca*.[65] There are, however, value differences in the classical economics treatment of the term gambling from the Hausa usage. The Hausa definition and perception of gambling as a great social vice is derived from Islam. Large amounts of Islamic literature including the Qur-an are unanimous in their condemnation of gambling.[66] In Islam, monies earned through gambling are considered profane. Because of this similarity or closeness in meaning between risk and gambling, the term 'risk' is often a pejorative, expressed with euphemisms. Generally, gamblers are perceived as people who lack self-control in their economic lifestyles. This gambling stereotype is extended to include very high-risk-assuming businesspeople.

Indigenous Hausa business operators are generally cautious about relatively unknown fields of economic pursuit. There is therefore a tendency to remain in familiar and safe, rewarding fields of commerce. Generally, risky endeavours are said to be stressful and human beings have a natural aversion towards stress (Rotter, 1966). Many indigenous entrepreneurs are of the opinion that economically risky activities or *cinikin kasada,* should be avoided as far as there are safer and equally profitable business alternatives available to pursue. Sometimes indigenous investors without Western education perceive manufacturing activity as an unnecessary gamble.

It is particularly interesting to note that most entrepreneurs in contemporary Kano are, to a large degree, uneducated at least in the Western sense. Two-thirds of the indigenous entrepreneurs interviewed in Kano had no formal education. Formal education here refers to Western education. Operating a modern manufacturing firm requires some form of Western education to fully understand the complexities involved. In handling the interview question relating to the reason for the low participation of the Kano business community in modern industry, a common answer received from respondents was 'lack of Western education' and the desire for 'safe investment'. An indigenous owner of a plastic-manufacturing firm covered by this study admitted that 'lack of Western education' has been his major handicap in running a modern factory. He further observed that this was the main reason 'many affluent and wealthy indigenous merchants avoided modern manufacturing ventures'. These limitations aside, indigenous entrepreneurs who successfully break away from commerce to industry are often considered exemplars of courage.

An analysis of risk perceptions among expatriate and indigenous business proprietors further reveals that, although most of them favoured risk in economic pursuits, there were variations about the extent to which it should be allowed. The Lebanese business proprietors strongly agreed that risk was inevitable in business activity, while 80 and 56 percent of their Chinese and Hausa business counterparts favoured risk-taking respectively. Apart from the culturally euphemistic connotation of risk itself, entrepreneurial caution tacitly reflects on the small size of many indigenous firms.[67] Operators of smaller firms sometimes try to minimise risk because of the fear that in case of losses their entire capital might dissipate. Since most successful entrepreneurs provide support to chains of dependants, or *barori*, a miscalculated risk on their part may throw several people into destitution. Therefore, indigenous entrepreneurs often have their families and large number of dependants to worry about.

Risk usually carries a time span. Investors are often interested in the period within which a task can be accomplished. The longer the period and the inability of an investor to decipher a profitable conclusion, the higher the tendency for reluctance. Investments in the manufacturing sector are best suited in a stable and predictable political climate. Conversely, business groups in Nigeria, and especially Kano, are rationally inclined towards commerce and fast profit-making ventures.

In nearly every society, there are 'cultural standards about what constitute appropriate or improper risk' (Douglas, 1985, p. 68). An individual's approach to risk may start from a culturally established norm of appropriateness. A group's shared culture provides a basic frame of reference on which areas or activities are risk-prone. This means that, in the face of uncertainty, most entrepreneurs would resort to axiomatisation of an idea of reasonable behaviour. Concerning various forms of culturally determined economic behaviour, Hannerz particularly observes that:

> The cultural entrepreneurs of the periphery carve out their own niche, find their own market segment, by developing a product more specifically attuned to the characteristics of their local consumers. The cultural businesses of the centre may have much greater material resources, but these local entrepreneurs have the advantage of knowing their territory. Their particular asset is cultural competence, cultural sensibility. Through their roots in local forms of life, they sense which concerns and tastes can be translated into market shares (Hannerz, 1995, pp. 44–45).

An essential trait of good entrepreneurship lies in the ability to decipher or predict with some amount of accuracy the taste of customers in a specific society or market. This ability to accurately perceive and adapt to the socioeconomic environment is described as 'induced preference' (Schoemaker, 1993). In many developing societies, entrepreneurial and managerial competence is measured by the ability to circumvent undesirable consequences. It is within this framework that variations in risk-related attitudes often emerge. Societal values shape people's behaviour when they are confronted with unknown situations. Entrepreneurial failures in business are therefore perceived as violations or contradictions of societal norms of reasonable behaviour. In view of this, natural disasters, such as fire outbreaks and rainstorms, are more easily tolerated than other forms of risk. The merchants whose failures are the products of natural disasters often attract more sympathy and support from their community and business associates than their experimentally unsuccessful counterparts.

Choice, or *zabi,* is an inseparable concept in risk analysis. The intertwining links among trust, choice, and risk makes their explanation teleological. Choice, which deals with preferences, 'applies logic to the act of choosing'. There is always a probability element in choice making, and 'the variance of the probability' is what constitutes risk (ibid.). Nearly every choice is beset with some form of uncertainty, in view of the alternative that is foregone. This is the reason most investors are cautious.

Rational business operators, whatever the economic considerations and strategy, would act according to group preferences and values, if they anticipate the survival of their businesses. Thus, rational choice is goal-oriented and involves 'efficient pursuit of preferences' (Stenmark, 1994). It is in this regard that the cultural values and norms of the group within which the businessman operates are relevant. When faced with a choice option involving economic investment, most individuals resort to a 'safety-first' strategy.[68] Safety is the degree to which risks are deemed acceptable by business operators and society (Ritchie, and Marshall, 1993, p. 104). Entrepreneurial desire for safety is one reason for investments in multiple business activities.

Multiple Investments as Safety Strategy

There has been in recent times a rising tide of multiple business activities in Kano and in other parts of Nigeria. Entrepreneurial or merchant desires to reduce risk and to ensure the safety of capital has given rise to multiple investments, in different and very often unrelated economic ventures. Merchants' quests for quick profits have been responsible for a multiplicity of investments in most African societies. Grégoire observes that:

> Although big business may absorb much of their attention, merchants will not pass up an opportunity to participate in some petty speculation that might produce a nice return. The richest *Alhazai* are those who have escaped from specialisation to diversity; they show a disposition and an aptitude to seize every opportunity to make money. This diversification, however, does not exclude favouring one category of activity or another (Grégoire, 1992, p. 146).

These risk-reduction strategies,[69] which are derivative of cultural priorities and persistently unstable political and socioeconomic factors, have combined to produce specific investment patterns and modes of rationality among indigenous and expatriate business communities in Kano. Many of the businesses currently operating in Kano are diversified businesses owned by Nigerian, Levantine, Asian, and European investors. Each of the leading Kano merchant industrialists, such as Gasash, Dan-Tata, Issiyaka Rabiu, etc., operates more than ten different businesses.[70]

The Hausa, Lebanese, and Chinese business proprietors on whom this study focuses have diversified business interests. However, the indigenous Hausa entrepreneur with seven business interests had the highest number of business interests. This is closely followed by the Chinese business investor, who operated six businesses, most of which were not closely related. My investigations revealed that the least diversified business interests were the four businesses operated by the Lebanese business proprietor.[71] However, this does not imply that Lebanese business interests in Kano are less diversified, the evidence suggests that they follow the same pattern.

Single individuals manifest entrepreneurial flexibility in the operation of a variety of smaller, highly labour-intensive firms. The relevance of flexible specialisation has been particularly stressed by Mendez-Rivero (1991) as a strategy for overcoming erratic demand whiles ensuring micro-enterprise development in developing countries. In contrast to this view, Marris and Sommerset (1971) have argued that indiscriminate diversification and the operation of multiple businesses are factors hindering the expansion and growth of African economies. Usually, profits earned from multiple investments are expended on farmland, houses, and additional spouses, which further weakens the capital base of small business operators. This trend, according to them, was particularly more widespread in commerce than in manufacturing activities.

Experience with the merchant industrialists of Kano has shown that business emphases during the year depend on the most thriving or prevailing economic activity in the society. This is usually done through the acquisition of information through business associates and well-acquainted customers. Fiet (1995) observes that the acquisition of risk-reducing information can be considered an investment involving some amount of cost. It is significant to note that in traditional societies where free information services are common, such costs are very small. The fear of natural disasters, such as fire outbreaks, which have become a perennial feature of the commercial life of Kano, has also encouraged multiple investments as a strategy for minimising total losses.

In expatriate companies, the risk-reduction techniques often employed include the nomination of influential persons and indigenous businesspeople into their business domains so those firms have greater recognition in the society. Usually, the expatriates give weak local partners, nicknamed 'sleeping directors,' board membership, which allows them full control and free access into the economy. The studies by Hoogvelt (1979) and Beckman (1982, pp. 37–51), on expatriate investment patterns in Kano are quite illuminating.

Risk perception has a correlation with changes in economic fortunes (Loewenstein, and Mather, 1990, pp. 155–175). Helson (1964) observes that in situations where the 'objective threat remains constant', risk declines. This is natural in the sense that, when individuals constantly encounter a particular type of threat, it becomes less relevant. However, this has not been the case with the Nigerian experience. Persistent political and economic instability implies that, rather than remain constant, risk threats to business activities continue to increase. Dahl (1998, p. 291) observes that without trust, the courage to face risk is low. Given the low level of trust among entrepreneurs, their risk-reduction strategies remain high.

3 Transition to a Modern Economy

> Practically every economic mechanism and institution known to us is found somewhere in the non-literate world . . . The distinctions to be drawn between literate and non-literate economies are consequently those of degree rather than of kind (Herskovits, 1952, p. 488).

This Chapter builds on the theoretical orientations of culture and entrepreneurship discussed in Chapters One and Two. The focus of analyses is on the way culture and religious reform pave the way for economic rationality among entrepreneurs in a northern Nigerian society centred in Kano. The first section of the Chapter, looks at the Islamic influence and the transition from a cash crop economy to a semi-industrial society. The second section of the Chapter focuses on selected aspects of Weber's work on the Islamic society as they relate to the empirical materials of the study. Much of the discussion in the final section of the Chapter however, deals with the resurgence of Muslim puritans, religious reform, and new forms of rationalisation in Hausa society.

The population of the city of Kano in northern Nigeria is overwhelmingly Muslim.[72] Census estimates in 1995 of over five and a half million inhabitants makes it the second largest city in Nigeria.[73] Despite the heterogeneous mix and a fair representation of all the country's ethnic groups, most of the people are Hausa or Hausawa. Among the many ethnic groups in northern Nigeria, the Hausa is the largest. The main cultural identity of the people is the hegemonic 'Islamic culture of the Sudan', with the Hausa language as the *lingua franca* of most inhabitants of Kano. The city of Kano, which was founded in the seventh century AD, is situated between the Sahelian zone, the savanna and forest regions of Central and Southern Nigeria. Peasant cultivation of export crops including groundnut and cotton, which led to economic prominence and the integration of Kano into the world economy, was facilitated by a favourable ecology. An average annual rainfall of between 700 and 800 millimetres, and a relatively fertile, light, sandy soil derived from sand dunes and loess deposits makes the region well suited for the cultivation of a variety of crops and animal husbandry.

Because of its economic and commercial significance within the sub-region, the city of Kano in northern Nigeria has been the focus of much research. Nonetheless, cross-cultural studies of enterprise behaviour in the organised private sector represent a narrow, recondite front. In most of these studies, entrepreneurial rationality and investment patterns are perceived as solely economic in origin. An in-depth examination of entrepreneurial patterns in Kano and elsewhere in the developing world reveals that they often violate traditional economic axioms. The economic and political significance of Kano lies in the fact that it is among the few cities in Nigeria whose prosperity preceded the oil boom (Tahir, 1975). Unlike many Nigerian cities, Kano has been the centre of the most radical and populist politics for many years (Olukoshi, 1988, p. 53). Kano's allure as an economically lucrative centre has attracted thousands of migrants. With the burgeoning demand for industrial goods in Kano, nearly seven out of ten houses by the roadside have a shop or outlet in which some commercial or business activity operates.

In Nigeria, Kano and commerce are cognate terms. With Kano business people, 'no profit margin is too small or too big' (Etuk, 1985).[74] The overwhelming preference for commerce is not fortuitous, but has its roots in over a thousand years of links to the trans-Saharan trade in which Kano had been a major commercial entrepôt of sub-Saharan Africa (Davidson, 1984). The merchant class[75] of Kano therefore has a long commercial pedigree. Like their business counterparts in other parts of Nigeria, the entrepreneurs of Kano show a proclivity for commerce, and portable assets rather than fixed investment in capital goods and modern manufacturing activities.

Most indigenous entrepreneurs will not participate in a business enterprise where personal supervision will not be possible, which implies a low trust of subordinates and partners in business, even if they are close siblings. The entrepreneurial desire to maintain a retinue of customers sometimes triggers inter- and intra-family competition. A new businessperson's friendly gestures to potential customers may be interpreted as meddling into personal affairs or encroaching upon another's business territory.[76] The struggle for economic supremacy apart, the successful entrepreneur will do everything in his way to stop other family members from 'falling too far' (op.cit., 1975, p. 101). This implicitly shows a desire for fairness in economic competition. Nevertheless, trust is essential for cooperation in modern business activities involving many participants. Economic progress partly demands the removal of barriers to cooperation and raising commitment beyond immediate family networks.

The growths of family businesses in Kano and parts of Nigeria are sometimes hindered by the widespread fear among their operators that expansion beyond family connections may lead to loss of control. Given the low level of interpersonal trust, single and small-scale enterprises dominate the business scene.[77] In Kano, the low level of interpersonal trust is mainly restricted to business dealings involving fellow Nigerians. While many local investors are often willing to do business with foreign investors, they have a strong aversion for joint business endeavours with fellow citizens. This implies an inverse trust relationship in economic transactions. Joint economic ventures among indigenous business operators are therefore few, despite the religious encouragement by the *ulama*.[78]

The selection of business partners in terms of their ethnic, social, and religious backgrounds requires considerable caution. Such precautions are usually products of unpalatable previous experiences in jointly operated businesses. In the manufacturing sector, an enormous division of labour is required to achieve the diverse technical, accounting and administrative skills necessary to man various departments. Such diverse requirements make direct supervision by business owners impossible. However, given the low level of trust for non-kin members in the control of their resources, entrepreneurial participation in manufacturing ventures compared to commerce is low.

The Islamic Influence

Durkheim (1961) and Berger (1967) represent two significant reactions concerning the role of religion to society. Whereas Durkheim stresses the integrative role of religion to society, Berger challenges this dominant position by arguing that religion is often an instrument of societal disintegration or destabilisation. Much of the evidence in Kano however, is consonant with Durkheim's view concerning the role of religion as an agent of societal integration.

Sources from the '*Kano Chronicle*',[79] suggest that Islam was introduced to Kano during the reign of Mohammed Rumfa in the second half of the fourteenth century by Wangara traders from the old empire of Mali.[80] Lubeck (1986, pp. 10) argues that Islam was probably introduced to Kano from a variety of sources including Mali, North Africa and Kanem-Borno. The spread of Islam in Hausaland is coterminous with the activities in the trans-Saharan trade. Early Arab and Tuareg traders from the northern parts of Africa were practising Muslims who diffused their religious faith among the local Hausa people.

Although coexisting and often overlapping in many areas, traditional Hausa and Islamic cultural values are not necessarily harmonious. In retrospect, friction between Islam and traditional institutions produced the early nineteenth century *jihad* in Hausaland led by Sheik Usman Dan Fodio. The reformist *jihad* had among its aims to consolidate on the gains already made by Islam, and additionally to enforce a practice of Islam without 'polytheistic associanism', apostasy, *bid'a*, or *shirk*. The successes of the *jihad* on the political, religious, and socioeconomic fronts culminated in the creation of emirates under the Sokoto caliphate. With Sokoto as religious and political headquarters, Kano became the caliphate's unrivalled economic centre.

Lubeck observes that the caliphate was the largest and most completely organised state system in West Africa. Some changes introduced by the *jihad* led to significant political and socioeconomic transformations in Hausaland. Traditional spiritual practices such as the *bori* cults and idol worshipping, considered antithetical to Islam, were reduced to the barest minimum. These limitations notwithstanding, Islam therefore provides the guiding principles by which most of the inhabitants of Kano conduct their daily affairs. Because of the influence of Islam, economic activities that contradict Islamic law carry social stigmas and, are therefore sometimes ignored, in spite of their profitability. Islamic prohibition of usury and aleatory contracts was one reason Muslim entrepreneurs resisted modern banking institutions during their early years of establishment in Kano and other parts of northern Nigeria.[81] Religious rejection of usury does not necessarily imply that Islamic economic doctrines inhibit business progress or successful entrepreneurship. They only imply that in Islam, economic practice is an integral and inseparable part of religious sanctity.[82] Rodinson argues that contrary to the explanations of most Western scholars:

> Islamic economic doctrines are essentially capitalistic, supporting and legislating private property, the pursuit of wealth and wage labour, honouring the honest merchant as a fighter in God's cause, sanctifying both profit and economic inequality as an expression of divine justice (Rodinson, 1980).

The impact of Islam is easily noticeable in Kano in the proliferation of mosques and places of Islamic worship. It is common to find in business premises and factory yards makeshift or permanent structures used as praying grounds by local entrepreneurs and their Muslim employees. Apart from congregating for five daily prayers, the mosque in the Islamic community is a forum for a diversity of other religious and social activities, which include naming and death ceremonies.[83] Man's imperative duty to Allah, with its requirement of continuous daily interactions in the mosque, provides avenues for strengthening communal fraternity. While the motive for congregating in a mosque may be essentially religious or social, daily interactions in religious worship between the rich and the poor often become avenues for establishing credit worthiness and corporate rapport. During certain times of the day, Kano has the semblance of Mecca, as most commercial and business activities come to a temporary halt for the afternoon prayers. However, once the prayers are over, the Muslim traders and factory workers will return to their economic activities without much ado.

Apart from the domineering role of Islam in Kano, there are sectarian divisions. The major Islamic groups include, the brotherhood, or *Sufis*,[84] *Sunnis*, *Shi'ites*, and *Yan Tatsine*. *Sufism* is generally associated with religious piety, self-denial, and the rejection of worldly or material things. It also implies 'mysticism', or *tasawuf*. The most prominent *Sufi* sects in Kano include the *Qadiriyya and Tijaniyya*.[85] Early works of Sufism and particularly the *Qadiriyya* are purported to have been introduced into Kano in the late fifteen century by Sheikh al Maghili (Paden, 1973, pp. 65–70). Historical sources suggest that the *Qadiriyya* brotherhood preceded the *Tijaniyya*, which was introduced into Kano during the early nineteenth century by Umar al-Futi (ibid.). Later reforms in the *Tijaniyya* however, came through religious contacts with Senegalese Muslim clerics and migrants. Certain distinct beliefs and practices distinguish the religious sects in Kano. Common practices among the *Tijaniyya* and *Qadiriyya* sects are the lengthy evening collective supplications, or *wazifa*, in which litanies are chanted in unison. In *Sufi*[86] religious networks, the evening *wazifa* sessions provide opportunities for strengthening communal relations between the poor, the learned *malams* and affluent merchant classes.

A combination of economic contradictions and social maladjustments in Kano led to the resurgence of a millenarian sect in late 1980, known as the *Yan Tatsine*. The emergence of this local variant of Islam marked the beginning of the radical shift in the religious and social life of Kano. Members of the *Yan Tatsine*, who lead esoteric lifestyles, are generally regarded as heretical in view of their muddled interpretations of the Qur-an and a militarist approach to religion. *Yan Tatsine* provocations in December 1980, led to the bloody Kano religious clashes in which over 8,000 lives were reported to have been lost. Subsequent *Yan Tatsine* insurrections in northern Nigeria include Maiduguri and Kaduna (October, 1982), Jimeta-Yola (February, 1984), and Gombe (April, 1985).[87] *Yan Tatsine* and *Shi'ite* riots in the 1980s were major threats to social stability and economic progress.

Perennial religious upheavals sometimes take a negative toll on human lives and business property. Government proscription of *Yan Tatsine* activities after the major Kano uprising in December 1980, inadvertently transformed the group into a secret cult. Though suppressed by an official ban, there are occasional outbursts of *Yan Tatsine* insurrections in the towns of northern Nigeria. Sometimes, the effects or consequences can be deleterious, given the nature of a prevailing social or religious provocation. Despite the religious differences and occasional misunderstandings, Muslim observance of the five pillars[88] of Islam provides common unifying ground for all the groups.

Religious Groups and Socioeconomic Networks

Intra-brotherhood solidarities in religion are sometimes extended to the economic sphere through the provision of interest-free loans, advancement of goods to new entrants in business, or the creation of sitting spaces in the vending booths and sheds of overcrowded markets. Religious solidarity also provides networks for introducing clients to foreign firms.[89] In particular, it plays a role in the circulation of economic and social support among business groups. This does not imply, however, that the local business entrepreneurs discriminate among their customers on the basis of religion or ethnic background. To a large extent, they are very open and flexible in their business dealings. Most of their social and economic activities are trans-ethnic in nature. Hausa traders, at home or as immigrants, are very tolerant and assimilate easily. Not surprisingly, therefore, their economic activities are noticeable in immigrant settlements otherwise known as *zango* throughout sub-Saharan Africa.

On the local front, some prosperous merchants and local industrialists are loyal students, or *muridai,* of the Kano *malams* and *ulama*. There are strong reciprocal ties among merchants, industrialists, the aristocrats, and religious leaders. These ties are sometimes manifest in marital bonds, fronting for the aristocrat in businesses, and merchants' financial support for the maintenance of sheiks' religious activities. Such acts of unsolicited generosity are often extended in anticipation of both Allah's reward and the sheiks' blessings for business success. Occasional investment by loyal merchants or *muridai* in the sheiks' social activities sometimes takes a toll on business profits. The merchant industrialists with the biggest business networks in Kano are those intimately involved in the activities of the *Sufi* brotherhood (Tahir, 1975; and Forrest, 1994).[90] Despite *Sufi* denunciation of worldly lifestyles, a combination of mercantile financial support and dual participation in commerce and scholarly activities puts its *sheiks* among the richest in Kano (op.cit., 1975). Nevertheless, merchant and business groups are the most distinguished classes in Hausa society.

The transformation of Kano from an essentially mercantile to a semi-industrial economy in the post-independence era has led to the emergence of an urban industrial working class in the city. Industrial workers and people who depend solely on wage employment for earning a livelihood are sometimes ranked among the urban poor. Because of the low wages paid to factory employees, accepting industrial work in Kano is akin to paying rents and being a perpetual debtor.[91] This scenario makes factory work less prestigious than commerce. Most factory workers, or *leburori,*[92] are practising Muslims. In view of this, employee upheavals in factories usually assume religious colourations, although this need not necessarily be the case. Lubeck (1986) further explains that Islam has become an aspect of working class identity, and is used as a vehicle of protest among the industrial workers. Nevertheless, the portrayal of Nigerian workers in class terms is persistently contradicted by a prevalence of non-class behaviour among them.[93]

Most of the tensions between expatriate industrialists and their employees are derivative of cross-cultural misunderstandings. Writing about the attitudes of a Hausa subordinate towards persons in higher authority, Tahir (1975, p. 100) observes that 'although he may genuflect before a superior and adopt an attitude of extreme humility and obsequiousness, this suddenly turns into a haughty self-assertive rebelliousness once the superior attempts to step out side the conventional forms to make a personal egotistical point'. Stepping outside conventional forms in work relations may include managerial rejection of an employee's request to attend the weekly congregational *jum'at* prayers or the denial of an employee the opportunity to attend the burial or *jana'iza* of a deceased neighbour. Extreme behaviour of this nature by expatriate industrialists often hastens tension and spontaneous strikes in the factories of Kano.

Daily practice of Islamic rituals in the factories by Muslim workers is another area that frequently generates mistrust and friction between them and their non-Muslim employers. Routinised observance of the five daily prayers by Muslim industrial workers is often considered by non-Muslim expatriate managers as an encroachment on business time. For Muslim employees, religious practice and the communion with Allah are ontological phenomena, and duty to Allah supersedes duty to any superior. These variations in personal philosophies or ideals between Muslim workers and expatriate employers sometimes engender friction in the factories of Kano. A common phrase which frequently punctuates the daily conversation of Hausa workers is the Arabic term *insha Allah*, or 'if God so wills'. Employee reference to Allah for all future assignments often conveys the erroneous impression to non-Muslim expatriates that they are lazy or unwilling to work. However, the evidence does suggest that most indigenous workers attach a great deal of importance to the jobs from which they earn their living. This implies that in dealing with indigenous Hausa employees however, the tenacious employer in Kano will often come to terms with the reality that is not only economic. Social and religious factors often matter in the work environment.

Towards a Conceptualisation of Islamic Sociology

In spite of the absence of a systematic study of Islam in sociology[94] Weber (1979) is crucial in providing the theoretical frontier and guidance on how culture or religion may influence rationality and economic behaviour in northern Nigeria. A scrupulously detailed analysis or stringent critique of Weber is not the aim this study, although it seeks to examine some aspects of his theory that has relevance to the empirical materials. Scholarly works featuring Weber and Islam are highly controversial for several reasons. First, Weber's presentation of the economy of Islam is more arcane than the rest of his work. Moreover, his planned work on Islam was not completed before his death (Bendix, 1962). This development has given rise to differing interpretations of Weber's work on the Islamic economy.[95] Although Weberian analyses are strong at the macro level, they can be shown to be weak at the micro-empirical level. In any case, scholars who often attempt to vindicate Weber, usually present his work as 'overlooked, underused or drastically misunderstood'.[96]

The problem, however, is not necessary a product of misunderstanding as of interpretive inconsistency of Weber when it comes to Islamic discourse. Thus, his failure to hold the same institutional prerequisite for rational capitalism constant, or to apply them in the study of China, India and Islamic societies makes him vulnerable to intense criticisms (Turner, 1974, p. 8). Given the analytical inconsistencies, 'no definitive or authoritative interpretation of Weber is genuinely possible' (ibid.). Interpretive understanding, or *verstehen,* and the formulation of meaning, or *sinn,* are central in Weberian methodology. According to Weber (1947), such understanding does not depend on empathy but on the analytical construction of the subjective category of actors. Thus, human behaviour must provide a meaning and capability of 'being understood in terms of subjective categories' (Peacock, 1978, p. 23). Categories shared by a plurality of actors in Weberian analysis are termed as cultural (ibid.). Actions become meaningful only when the full motivations of actors are properly understood (Weber, 1966, p. 14). For Turner (op.cit.), Weber's inability to stick to personal rules makes him vulnerable to criticism. Throughout his analysis of Islam, 'he fails to hold to one consistent position, slipping between interactionism *verstehende* sociology and forms of determinism'.

Early Christian theology and Western literature have over the years presented a distorted view of Islam as a violent religion. In much of this, Islam conjures a picture of a marching band of religious fanatics with savage beards and fiery eyes brandishing drawn swords and attacking the infidels (Maududi, 1983, p. 1). Today, some negative themes regarding religious intolerance, fanaticism, hedonism, and pervasive subjugation of women in Islam are still prevalent. In particular, the Muslim warrior[97] is given prominence throughout

Weber's analysis of Islam. The social role of the Muslim warrior was completely overstated because Weber was probably unaware of the merchant class in shaping the values of Islam (Turner, 1974; and Rodinson, 1980).

It is worth noting that the spread of Islam in northern Nigeria as in many Islamic societies is coterminous with the activities between indigenous and Arab merchants in the trans-Saharan trade. That Muslim 'warriors were the main social carriers of the Islamic faith' is not only controversial but refutable in the northern Nigerian context.

History reveals that Indonesia, the most populated Islamic country in the world, never experienced Muslim invasion or Arab conquest. Similarly Malaysia and many countries with overwhelming Muslim populations do not have a history of Islamic invasion (Deedat, 1992). A larger part of the Muslim population on the African continent except perhaps for the coastal fringes of North Africa, got converted into Islam through contact with caravan merchants during the 8th and 9th centuries (Davidson, 1984). The fundamental reasons behind the rapid spread of Islam as observed by Deedat (1992) include, the attractiveness of its universal doctrines of the brotherhood of mankind, abolition of the caste system, the right of women to inherit, humility and simple life of its founder Muhammad (S.A.W.), the idea of unity of God which was easily understood, and prohibition of intoxicants.

Equally important in Weber's analysis regarding Islam is the assertion that rational law, as well as autonomous economic and political stability, which are *sine qua non* conditions for development, were absent in Islam. Political stability was absent according to him because of the disruptive role of the Muslim warrior. Weber thus, characterises the Muslim economy as patrimonial and dominated by prebendal feudalism (op.cit., 1974, p. 8). Personalised traditional power is described as 'patrimonialism', and it is distinguished from modern law, which is depersonalised, and 'bureaucratic'.[98] Reacting to the latter, Amin (1994, p. 324) posits that the word patrimonialism has been severely abused in Weberian formulation because the so-called 'patrimonial' system in no way disregarded the law. In contrast, the hallmark of Muslim society is both law, or *sharia,* and reason. The *sharia* or Islamic legal framework is a legal instrument for settling disputes, assessing inheritance property in northern Nigeria.

Amin criticises Weber's thesis as 'very Germanic in trying to stretch German history to cover the whole of mankind'. The application of this 'dubious' historical thesis to the specific characteristics and personalisation of power in Africa is therefore inadequate, according to Amin (1980). The principal element of tributary ideology, he notes, is not 'patrimonialism' but 'metaphysical domination' through the autonomous power of religious institutions, which often compensate for the state's failure. Conversely, Confucian China, Islam in the Caliphate, or post-colonial power in African countries cannot be described as patrimonial. Accordingly, the African ideologies of kinship and communal stage of development are more appropriate in characterising African societies rather than patrimonial domination (ibid.). Kinship and ethnic networks dominate not only in the domain of political power, but in the economy as well.

In attempting to show the correspondence between the Protestant religion, merchant capitalism, and modern forms of law of philosophical thought, Weber (1958) considered the economies of especially Muslim and Asiatic societies non-rational. The political and economic conditions of oriental societies in his thesis are identified as hostile and incompatible with capitalist prerequisites. Despite Weber's claim that individual prerequisites for rational capitalism were present in other civilisations, he however, rejects the notion that such conditions were present in India, China, and the Islamic world of the Middle East (Turner, 1974, p. 12). Weber's analysis of the rise of capitalism in modern Europe, however, fails to prove whether the religious features preceded or were created by the economic regimes (Rodinson, 1980). In Weberian (1980), thesis therefore, Western civilisation is distinguished 'from every other' in its rational conduct of life. It is necessary to note, however, that the analysis regarding capitalism and Protestantism has geo-historical and cultural limitations.

Counterpoising the view on rationality, Herskovits (1952, p. 488) assails that the mechanisms and institutions in occidental and oriental societies are similar and economic differences are 'those of a degree rather than a kind'. In dealing with the question of rationality therefore, Weber becomes very ideological by using Western society as the absolute centre of reference. Weber's (1974) distinction between substantive and formal rationality pays less attention to cultural change resulting from modernity. Although modernity may have other 'sombre' sides, its 'opportunity sides' was stressed most strongly by the founders of sociology (Giddens, 1995, p. 7). Weber was the most pessimistic among the founding fathers partly because 'he did not fully anticipate how extensive . . . modernity would turn out to be' (ibid.). Although Weber was of the view that history moves towards rationalisation, he does not provide explicit explanations regarding the future of occidental economies.

Human diversity, however, suggests that many kinds of cultural response to modernity are possible (Giddens, 1995, p. 175). Varieties of cultural reaction to modernity have resulted in the transformation of economic regimes and societies until now considered stagnant and unproductive into 'tigers', and 'Muslim puritans'.[99] Despite the current economic turmoil in Southeast Asia, significant economic achievements in this region over the past three decades have undoubtedly shattered basic clichés of Weberian theses regarding economic performance in occidental societies. Today, the so called third world countries in the Islamic world, including Kuwait, Saudi Arabia, Bahrain, and Qatar enjoy higher per capita incomes than most industrially advanced countries (Seers, 1977).[100]

Studies of dual economies, including Kano, show that rationality is subject to continual corrections and adjustments in the course of history and tradition. Rapid transformations of the Nigerian society over the years has led to the introduction, practice and acceptance of abstract systems. Nowadays, modern financial institutions and industrial activities are widely embraced by the indigenous entrepreneurs. This seems to suggest that rationality or reasonable behaviour depends on time and place.[101] As noted earlier, rationality is not fixed, it keeps changing and adjusting with the passing of time and history. Thus, what constituted rationality in the precapitalist or medieval era may be seen as irrationality in the contemporary era. Thus, any economic behaviour that seems irrational may be found to possess some rationality of its own when placed within the overall functioning of society (Godelier, 1972, p. 317). There is no rationality 'in itself', nor any absolute rationality. Godelier observes that:

> What is rational today may be irrational tomorrow, what is rational in one society may be irrational in another. Finally, there is no exclusive economic rationality . . . In the end, the idea of rationality obliges us to analyse the basis of structures of social life, their *raison d'être* and their evolution . . . Synchronic and diachronic analysis of past and present social systems would enable us to get an inkling of the 'possibilities' of evolution inherent in these systems (Godelier, 1972, p. 317).

For Weber, rational bureaucracy accounted for the development of Western societies. The bureaucratic organisation offered effective control of nature and society, liberated man from the anxieties and domination of magical forces.[102] In spite of the advantages which bureaucracy offers to modern society, it has also created pervasive social problems in terms of alienation and depression arising from the highly individualistic nature of Western society.

Experience in Kano suggests that, in a dual economy where pre-capitalist modes of production and the capitalist modern industry coexist, varieties of rationalities are possible. On the one hand is the capitalist modern industry of the formal sector with its bureaucratic requirements; on the other is the informal sector with traditional or pre-capitalist non-bureaucratic commercial networks. Both, however, demonstrate a proclivity towards economic rationality and progress. That traditional Hausa merchants in commerce are less bureaucratic does not imply that they are dominated by magical forces as will seem to have been posited in Weberian thesis. The truism is that magic, or *sihr*, and Islam are incompatible in contemporary Kano.

The Reconstruction of Rationality

The concept of rationality has become a complex analytical tool in view of the multiple referents. That rationality is complex, however, does not imply a complete vagueness. To understand the economic rationality of a particular system, as Godelier suggests, it is necessary to analyse the rationality of economic agents who are found in the system. In particular, it is necessary to reduce the category[103] into two specific agents, entrepreneurs, and workers. Entrepreneurs here refer to traders, and industrialists and their agents or managers in the economic system of northern Nigeria. The employee or worker category includes manual and non-manual workers of both the formal and informal sectors of the system. I do not include consumers because both entrepreneurs and workers fall under this category. Differences of inequalities between workers and entrepreneurs in the control of capital imply that the manners in which they rationalise their activities and pursue interests are different. Although wages may encroach on entrepreneurial profits, employee rational behaviour and compliance of organisational rules are always geared towards increased wages and salaries.

Economic rationality has several meanings and applications and may refer to any of the following: effectiveness, efficiency, profitability, return, productivity, minimisation of cost, maximisation of profit, maximisation of satisfaction, optimum decision, choice, calculation, forecasting, management and organisation of labour, of an enterprise, of a branch of industry, of national economy, development, balance growth, progress, distribution, justice etc. (Godelier, 1972, p. 9). Given these conditions, the rational entrepreneur, regardless of education and cultural background, will employ every available strategy to seek effectiveness, efficiency, profitability, and higher returns on his or her investment.

Rationality is not only ' . . . about alternatives of means-ends causality, and judging the relative cost of alternative means-ends avenues, but also to acquire the most realistic notions of alternatives . . . so as to allow an indication and choice of goals to be sought' or pitfalls 'to be avoided' (Himmelstrand, 1995, p. 4). According to Maurice Allais (1955, p. 33), a person may be considered rational when he/she pursues ends that are mutually coherent, and he/she 'employs means that are appropriate to the ends pursued'. The activities pursued by entrepreneurs and workers only become rational if the ends are mutually coherent and the means employed are appropriate to these ends. The rationality of entrepreneurs and workers are expressed in the activities of production and distribution.

An entrepreneur is rational if he combines personal income or capital in a calculable manner that ensures the generation of surplus or profit. Maximisation of profit is a rational behaviour of the entrepreneur. Both entrepreneurs and their workers strive towards profit acquisition with the foreknowledge that it provides the fulcrum for their enduring partnership. Entrepreneurial rational decision takes the form of risk calculation and encourages the selection of appropriate strategies that will yield results. Entrepreneurial rationality is incremental in the sense that it seeks to achieve increases in surplus. For the entrepreneur, it implies good management and profitability. For the employee or worker, it implies placing oneself under management, as well as contributing toward the realisation of profits (Godelier, 1972, p. 47). The rational employee therefore participates 'actively and totally in the functioning of the enterprise, and identifies his interest in the quest towards profitability. Furthermore, a rational worker or employee is one who does his work according to the 'best norms', by economising both the time and, in turn, the money of his enterprise (ibid. p. 36).

Rationality does not exist in a vacuum. It is backed by the behaviour of actors in an economic regime and the capacity of such a system to ensure growth of the means of production and improvement in the standard of living (Godelier, 1972, p. 11). From a theoretical perspective, it deals with conditions of attaining certain objectives, while taking into account certain environmental constraints (ibid. p. 12). Societal change is contingent upon the compatibility or incompatibility of unintentional properties of inner structures composing the systems. Rationality leads to ineluctable assertions often supported by proofs of the superiority of one economic over the other, and of the necessity for one of them to vanquish or disappear during historical development. Nevertheless, history shows that more often than not, economies do not easily disappear. Rather they adopt or assume new shapes and characteristics.

Rationality is influenced by the environment and its constraints. Environmental structures and constraints eventually influence the strategies, final or eventual outcome of the options or measures which entrepreneurs adopt in a society. Some of these constraints also include logistics, past history of business performance, the reliability of the legal system in maintaining contract law, the government, economic policies, such as tax regimes and foreign exchange allocations and interest rates on venture capital.

To understand the rationality of economic behaviour of individuals, it is necessary to know the structures of the economic and social systems within which they operate (Godelier, 1972, p. 26). As such, generalisations, or jejune categorisation of societies as rational, and non-rational economies can very be misleading if they do not take cognisance of internal structures or constraints, and of the performance actors within such structures. People are just not rational for the mere sake of it. They are rational because they weigh alternative options and anticipate consequences for their action. Instrumental rationality postulates of neoclassical theory assumes that actors possess relevant information necessary for correct evaluation of alternatives and in the process make choices that will achieve the desired ends (North, 1996, p. 108).

Rational choice theories often ignore the crucial question of timely access to accurate information. Imperfect communication networks and delayed access to information on market conditions especially in developing countries often result in speculative business and untold losses to business operators. Entrepreneurial decisions and investments in multiple economic ventures are not taken fortuitously but in recognition of the environmental challenges and constraints. That most business people are less inclined towards joint business projects is not necessarily a product of irrationality. Previous unpalatable experiences in joint ventures are responsible such behaviours. Considering many factors, it will be fair to assert that the local entrepreneurs are rational in their economic pursuits. The spread of investments and low patronage of insurance firms is exacerbated by the lack of reliability and poor performance of these firms.

Given the unstable nature of Nigeria's political economy, the investment patterns of indigenous entrepreneurs in Kano and across northern Nigeria suggest that they are rational. Rational calculation is an essential part of the entrepreneurial process in every society. This process, which is the starting point of capital formation, accumulation, and expansion, is concerned with the innovative behaviour of the individual. Thus, the entrepreneurial process, for Dobb (1963, p. 218), is primarily concerned with the exploitation of production advantages such as cost reduction, technical improvement, and optimal utilisation of labour.

In the commercial sector, the merchant practice of keeping large sums of money at home or in shops does not necessarily imply that they are irrational. The economic demands of the business sector in many developing societies including northern Nigeria, sometimes demands the availability of cash at all times. Most commercial banks close on the weekends, and the only means by which local entrepreneurs can take advantage of occasional windfalls of cheaper commodities and 'hot-goods' is by having cash available around the clock. This suggests that the rationality or irrationality of entrepreneurial action depends on the environmental context within which it occurs.

Muslim Puritans and Islamic Reform

The most significant religious reform in Hausaland came during the eighteenth Century Fulani led jihad. The Jihad was reformist in the sense that it regarded Islamic practices in Hausaland as 'backsliding and syncretistic' (Paden, 1973, p. 45). Paden identifies four distinct usages with the idea of reform in Kano. Accordingly, reform includes 'purification, modernisation, westernisation, and secularisation'. Several years after Paden's study, the term 'secularisation', often associated with separation of religion from politics and other public matters, has assumed a more pejorative and anti-Islamic character in northern Nigeria. As such, we do not consider it consonant with the concept of reform in contemporary Kano. For many local observers, westernisation, which means the adoption of western lifestyles by locally educated groups, is a twin sister of secularisation. In the Kano context, secularisation and westernisation are inseparable because the two need each other to survive. In order to avoid this pitfall, we conceptualise reform as a marriage between religious purification and modernity. The Qur-an and Hadith are essentially the reformists' guiding principles. Modernisation or modernity however, entails, mass participation, scale expansion, use of modern technology, and universalistic criteria of membership inclusion (ibid. p. 70).

Part of the reform in the *Qadiriyya* and *Tijaniyya* brotherhood was the anti-colonial militarism in the early 1950s in northern Nigeria. Colonialism not only posed cultural and political threat to reformers it was also seen as an economic monopoly of the cash crop exports. Reformist resistance to colonialism therefore has cultural, political, and economic agendas. Tahir (1975) documents *ulama* and reformists encouragement of business groups to imbibe a culture of investments in joint businesses. By encouraging partnership ventures among Muslim business groups, the reformers also encouraged the building of trust bonds in the Hausa business community.

Reform is a dynamic process in the sense that, anytime society backslides, reformers see upon themselves having a duty to ensure a return to true Islam. In 1978, a new wave of Islamic evangelism and the desire among some Muslim factions to eradicate local innovativeness, or *bid'a*, led to the formation of the *Izalat al Bid'a wa Iqamat al Sunnah*. The *Izala*, which is reformist and mainly *Sunni*, advocates a 'return to Source'.[104] *Sunni* doctrines stress the Unitarianism of God or *wahdaniyya*. In particular, *Sufi* beliefs and ceremonial practices are strongly criticised by the *Izala* and Orthodox *Sunni* Muslims as syncretic, *bid'a,* or *shirk*. Usually, such condemnations are backed by Quranic verses and *Hadith* which are antithetical to the notions of intercession and associanism.[105] The local resistance to the *Izala* movement during its formative years was due to its radical condemnation of religious syncretism. Sometimes, attempts to proselytise non-sectarian members have resulted in violent clashes between the groups. *Sunni* adherents are stricter in the enforcement of Islamic tenets, they are also generally considered more flexible towards commerce, science, and modernity than *Sufis*.

The most fanatical and intolerant groups concerning social, economic and political matters, are found among *Shi'ites* or *Yan Shi'a.* Fundamentalist *Shi'ites,* also known as Muslim Brothers, or *Ikhwan,* are particularly of the opinion that economic earnings can only be considered pure under an Islamic state. The call among *Shi'ites* for the establishment of a Muslim state in Nigeria under *sharia* law has led to frequent clashes between enforcement agents and local *Shi'ite* groups. In recent years, religious extremism within the group has led to speculations about Iranian or pro-Iranian influence on religious activities in northern Nigeria.

Islamic *sharia* and the four major legal schools[106] support private property. In Islam, there is no limit to property acquisition while such property is the product of genuine effort. Islamic economic doctrines, therefore, are not obstacles to wealth acquisition. Islam also enjoins transparency and honesty while rejecting all business dealings involving aleatory terms (Imam-Malik, Chapter 31, pp. 247–278; and Al-Bukhari, Vol. 3, Chapters, 34–37). Islamic economic prohibitions are essentially directed against certain practices that inhibit the free working of the liberal economy (Rodinson, 1980, p. 17). In Hausa business networks, moral precepts are applied in the ranking of people.

Cultural and religious ethics require that the exemplary businessman or woman must possess the qualities of truthfulness, or *gaskiya*, trustworthiness, or *amana*, generosity, or *karamci*, patience, or *hakuri,* and sobriety, or *hankali* (Tahir, 1975, p. 140). Above all, the exemplary business person has to be collectivist rather than individualistic in his/her economic pursuits. Islamic ethics make it obligatory for the healthy Muslim to seek occupation and treats work as an essential part of faith. These cultural values and fundamental beliefs are not antithetical but positively encourage societal progress. In Islamic communities, the respected man is one who is always busy doing something, no matter how trivial it might be.[107] Islamic doctrines, therefore, encourage neither idleness nor laziness. Rodison (1980) argues that by encouraging the accumulation of wealth and personal property, Islamic economic institutions are essentially capitalistic.

The Islamisation of Kano stimulated certain commercial activities, notable among them the production and marketing of rams for Islamic rituals. The prohibition of alcoholic consumption by Islam has resulted in grain surpluses and an extensive grain market. Islam has given a boost to the textile market because of the need for textiles to make Islamic gowns and dresses according to Islamic norms of modesty (Lubeck, 1986). Hausa societal respect for heroism and the application of economic success as a criterion in the ranking of men further encourages individual pursuit of economic distinction. This tacitly implies that, although aristocratic status is inherited, success in other fields of human endeavour, including business, can catapult a commoner, or *talaka*, into the aristocratic ranks.[108]

Although Weber regards Islam as 'the polar opposite of Puritanism',[109] I consider them inseparable. Furthermore, his prerequisites for Puritanism are present in Islam for three basic reasons:

1. In Islam, religion and economic pursuits are inseparable. This is because wealth must have a religious sanctity to be accorded proper recognition. Islamic economic ethics positively encourage the free working of the liberal economy without placing limits to wealth acquisition.
2. Contrary to Weber's view, hedonism, which inhibits the accumulation of capital, is discouraged by Islam. The Qur-an, a sacrosanct book of Muslims, denounces hedonism in its portrayal of 'spendthrifts as companions of Satan'.
3. The quest for religious purity has continuously led Muslim reformers and new jihadists to revoke traditional practices regarded as syncretic and antithetical to the ideals of pure Islam.

Today, Muslim puritans in Kano and parts of northern Nigeria have come to accept that hard work and piety must supersede superstitious beliefs in business transaction. In view of this, indigenous Muslim entrepreneurs have revoked mystical and magical practices, and are today less dependant on the potency or efficacy of charms, which had led to reckless use of venture capital and the collapse of businesses. Thus, through the religious rituals, or *ibada*, and methodological adherence to norms that they continue to rationalise into a strict ethical code, Muslim puritans are able to suppress sensuous impulses, which are not consonant with the demands of modernity.[110] Thus, to rationalise his life, the purist amputated or revoked tradition (Peacock, 1978, p. 22). The relationships between reform and rationalisation as postulated by Weber are therefore present in parts of northern Nigeria and especially among the Muslim business groups of Kano.

Yet, I agree with Peacock (1972) that the Puritanism of Muslim business groups has 'a distinctively Islamic colouration'. Unlike Protestantism which is 'individualistic and dramatistic', Islamic rationalisation is 'collectivist and legalistic' (ibid.).The Islamic tithe, or *zakkat,* provides the legal framework for measuring or assessing economic transactions. Collectivism as a form of rationalisation implies that the successful entrepreneur in the Muslim Hausa must be 'his brother's keeper'. As such, entrepreneurial surplus earnings are often shared among the local community as *zakkat* or *sadakat*. In the Islamic society, paying *zakkat* is a way of purifying wealth. Consequently, the successful merchant who sanctifies profit and uses it judiciously fights in God's course. In view of this, a sympathetic affluent merchant in Hausa society is regarded as a blessing to an impoverished community. Most often, the less-fortunate members of society including beggars, the handicapped, and poor Islamic scholars must live on the support of the benevolent merchants.

With religious reform, many Muslim entrepreneurs are becoming increasingly aware of the religious implications of their economic activities. More importantly is the widespread belief that wealth, if properly handled, could pave the way for the life beyond. This privileged position makes merchant classes distinct figures in Hausa society. Like their puritan counterparts, Muslim merchants created a calling in the world which held business motives in the highest regard.

Puritan ethics require entrepreneurial asceticism and shrewd management of resources a prerequisite of rationalisation. Consequently, a 'person who lives as a worldly ascetic is rationalist', not only for the systematisation of 'his own conduct, but for the rejection of everything that is ethically irrational' (Weber, 1978, p. 544). It therefore seems that asceticism is a relevant condition for business merit. Worldly asceticism according to Weber, was the instrument of proving religious merit among ascetic Protestants.

In contrast, Islamic economic attitudes were presented by Weber as essentially hedonistic. He argues that 'whereas Puritan ethical motive encouraged capital investment in the west, in Islam it was bound up in personal commodities and squandered' (Turner, 1974). The practice of polygamy in Islam was particularly stressed as a reflection of profligate and hedonistic consumption patterns. Such a critique, however, gives no recourse to the numerical prominence of women in the world community and the historical context and rules regulating the practice of polygamy in Islam.

The Muslim world-view regarding asceticism, or *zuhd*, is relevant for understanding the philosophy of business groups in Muslim communities. For Kinberg (1986, pp. 27–44), *zuhd* refers to the renunciation and suppression of evil inclinations. Its central essence is the purification of the soul and total devotion to God. Devotion to God requires effort in this world on behalf of the world beyond. Thus, the *zahid*, or ascetic, is not to deny the fact that he lives in the world but that he has to deal with it properly. Dealing with the world properly extends beyond the economic into the spiritual sphere where the individual seeks daily sustenance. A look at successful entrepreneurs in northern

Nigeria and elsewhere in the world suggests that modest beginnings characterised by initial temperance, parsimony, or self-denial are crucial to business progress. Entrepreneurial self-denial or asceticism, as illustrated, has both economic and spiritual dimensions.

Islamic Reform and Rationalisation

For rationalisation of action to proceed in the Weberian context, there must be spiritual change capable of rendering the process meaningful and legitimate to the actors involved. A typical example was the reformation movement of Protestantism, which culminated to Calvinism of the sixteenth and seventeenth-century Puritans (Peacock, 1978, p. 22). Thus rationalisation for the Puritan 'of the conduct within the world, but for the sake of the world beyond' (Weber, 1958, p. 154). There is a logical sequence between economic rationalisation and religious reform. Among the Muslim puritans of Kano and parts of northern Nigeria, success on earth is often perceived as signal for preordained success in the hereafter. Not surprisingly, daily Hausa aphorisms positively encourage self-distinction for the sake of life beyond. Failure is dreaded, not for the shame and worthlessness it brings to the individual, but for feeling that it could imply failure in the more permanent world beyond. Thus, a man who labours to feed his family is not only saving them from hunger but also fulfilling a rational religious duty.

The eighteenth-century reformist jihad in Hausaland provided avenues through religious networks with the Arab world, and stimulated commercial links across the Sahara, which eventually saw the emergence of local merchant classes. Similarly, it introduced literacy and new ways of communication hitherto unknown in Hausaland. The adoption of a new written form known as *ajami* by the sixteenth century was particularly significant, in the sense that it led to the introduction of new forms of rationalisation in Hausa society. In particular, it eased local governance and provided a legal basis for economic transaction. Commercial earnings became easily calculable and credit allocations more legally binding. Literal knowledge of *ajami*, facilitated communication among the merchant classes in the commercial nexus. Improved access to communication, among other things, reduced the cost of transaction in the economic realm. With literal knowledge and ability to communicate in written form, the need for face to face transaction became less likely among entrepreneurs.

Islamic reform among the brotherhood in the early 1950s encouraged the adoption of the modern industry and any aspects of modernity that did not pose a threat or inhibit Islamic religious practice. The result of this was significant diversification from commerce and cash crop exports to modern industry. With less than ten indigenous factories in 1950, there were over 58 indigenous factories in Kano by the late 1970s. Most of these factories today, manufacture textiles, plastic, insecticide, galvanised iron sheets, and process food. There are now over 100 modern manufacturing firms belonging exclusively to indigenous entrepreneurs within the Kano metropolis and its environs. That local entrepreneurs have accepted to invest in modern industry and other economic activities without external pressure or guidance shows that they are rational in their quest for increased productivity and profit. The evidence in Kano shows that pure Islam rejects neither modernisation nor industrialisation. Not surprisingly, some leading Islamic scholars in Kano and parts of northern Nigeria are also among the leading merchants and industrialists.[111]

Evidence shows that puritanical Islam encourages economic progress and development in the Islamic society. Through 'rational pursuit of self-interest', the entrepreneurs of Kano in northern Nigeria have gradually reconstructed society from a bazaar and peasant oriented cash crop economy to semi-industrial society.[112] There is a ubiquitous presence of the 'spirit of enterprise' and the incentive to invest among indigenous entrepreneurs a basic condition for development. Although trust among business groups is generally low, religious reformers or new jihadists in economic activities are less likely to renege on their promises than syncretists. Thus, 'the reformist vision is more collectivist than private' perceiving personal 'contributions to the wider society, religion and ethnic group rather personal and domestic happiness' (Peacock, 1978, pp. 203).

In recent years, the Muslim entrepreneurs of northern Nigeria and parts of the developing world have become increasingly rational in their economic pursuits, due to rapid socioeconomic transformations, of modernisation, industrialisation, and globalisation processes. There is also ample evidence in northern Nigeria and parts of the Islamic world to show that Islam and modernity can coexist in a stable equilibrium.

In spite of the current spiraling socioeconomic and political turmoil in Nigeria, indigenous cultures are generally receptive to modernity and change. In the Islamised parts of northern Nigeria, the daily practices of the people show that those aspects of modernity which are antithetical to Islamic teachings and practices are essentially of a moral and ethical nature. Developmental changes are therefore acceptable in Islamic societies if they do not interfere with religious faith and practices.[113] In Hausa society, development is generally held as divine assistance through men in the improvement of living conditions on earth. The agents of development are therefore Allah's instruments for change.

Change is a necessary component of progress. A Hausa aphorism that 'a change in the tempo of the drums requires a proportionate change in dancing steps', underscores the relevance of attitudinal changes to the ever-changing world. This implies that societal attitudes about food production have to shift from subsistence to mass food production levels capable of sustaining large populations and providing raw materials to the modern industrial sector. Commenting on African cultural receptiveness to change and development, Ikiara observes that:

> A notable advantage present in the African society is the absence of serious institutional and ideological barriers to modernisation that are characteristic of some parts of the developing world. Africa has a strong egalitarian tradition that does not encompass a hierarchical division of society. This is an important attribute in development as the entire society is open to innovations (Ikiara, 1994, pp. 126).

In analysing cultural tolerance of change in the Hausa society, Callaway (1987, p. 211) observes that it 'has historically proved to be very adaptable to outside forces as witnessed by its assimilation of Fulani rule, in incorporation of Islam, and accommodation of British colonialism'.

Contrary to widely held notions elsewhere, modernity and development are complementary to Islam in several ways. It is now generally agreed that science is the epitome of modernity. The progress in modern science has resulted in the ability to fight and control diseases in most Islamic societies, including Kano. Scientific achievements in medicine complement religion in the sense that good health is an essential component of religious worship. The vaccination of millions of Muslims before *hajj* has now become the safest means of preventing the spread of large-scale epidemics during the annual pilgrimage.

Furthermore, aviation transport makes a journey between Kano and Mecca feasible within a couple of hours. In the years before the invention of the airplane, such a journey was a lifetime undertaking that often took several years of trekking through unknown terrain. Many pilgrims ended up in captivity as slaves. The use of industrially manufactured clocks and wrist watches allow Muslims to know the exact praying times or when to break a fast, even in poor weather conditions. With industrial loudspeakers in mosques, muezzin calls for prayers can be heard across several kilometres.

Today, modernity to the local people implies the availability of potable water, a good road network, easy access to transport facilities, hospitals, schools, and industrial jobs for the ever-growing labour market. The introduction of modern publishing and copying equipment has given a boost to diverse literary and scholarly activities in Kano and other parts of northern Nigeria. Similarly, the modern manufacturing sector continues to offer employment to hundreds of thousands of people in Kano and other parts of Nigeria. The increasing demand for industrial goods has boosted the activity of firms in the modern manufacturing sector in the post-independence era. The evidence suggests that the Muslim Hausa society in northern Nigeria is generally receptive to changes that complement societal living standards. This conjecture, however, does not foreclose the challenges and obstacles that are sometimes posed by obscurantist beliefs and local reactionary religious interpretations.[114] Nevertheless, modernisation entails change and new forms of rationalisation in society. Many developing societies, including northern Nigeria, are becoming more rational in view of local reforms necessitated by modernity.

4 The Concept of Trust

> There are obviously some cases which call for trust and other cases which call for distrust. This rationality is indisputable . . . In ethics, it must therefore be assumed that whether in a particular instance one should trust, or not, follows from the objective features of the situation, from common human understanding. Therefore, even if there is a formulated general rule about trust as principle, the decision as to whether it should be followed or not must be delegated and left to the situation (Luhmann, 1979, p. 86).

In Chapter Two, I discussed risk-related attitudes and multiple investment patterns among the entrepreneurs of northern Nigeria. I propose to show in this Chapter that trust, which operates to facilitate the reduction of risk, influences entrepreneurial business decisions, and investment strategies as well. Without trust, the courage to face risk would be weakened (Dahl, 1998, p. 291). In particular, the low level of trust among entrepreneurs in the economic sphere is a major factor behind multiple investment attitudes. Risk and uncertainty are the two necessary preconditions for trust (Coleman, 1990; Bhide and Stevenson, 1991). Risk and trust are intertwined.

Given the unpredictable nature of the Nigerian economic climate, the safety strategy among business groups is the dispersal of capital into multiple business ventures. Multiple investment in unrelated economic ventures ensures the spread of risk and the provision of security against unforeseen contingencies. For Giddens (1995, p. 148), 'trust and risk, opportunity and danger are polar, paradoxical features of modernity' which 'permeate all aspects of day-to-day life'. It is, however, unclear whether risk precedes trust or is a product of it. Trust, which refers to confidence in and reliability of a person or system, is examined within a business context. A trust relationship is characterised by consistency, honesty, integrity, commitment, and responsibility.[115] From a sociological standpoint, trust is a basic requirement in the daily relations that characterise normal societal life (Luhmann, 1979).

In its cultural manifestation, trust symbolises virtue and therefore has a moralistic value. Trust, as Fukuyama (1996) observes, matters because it has a cultural value that relates to economic prosperity. In the political forum, a modicum of trust is necessary for the proper functioning of the activities of different groups that often make up a nation (Lagerspetz, 1992a). Cultural and religious values often provide moral and practical interpretations on what or who can be trusted in a given system or environment. Generally, 'religious beliefs' tend to 'inject reliability into the experience of events or situations' (Giddens, 1995, p. 103). Islam, which literally implies surrendering faith to the almighty Allah, is built on a principle of reciprocal trust between human beings and their creator.[116]

Four sources of trust or *amana* can be recognised in Hausa society. First, trust is derived from Islamic religious ethics. In Islam, the individual is accountable for a breach of trust both in life and after death. The dual relevance of trust in society lies in its use for building bonds and settling disputes. Islamic religious teachings and Hausa cultural ethics encourage the upholding of trust among individuals and groups. Trust in God is exhibited in the religious and social spheres of life. The five daily prayers, fasting during the month of *Ramadan*, payments of the *zakkat*, and embarking upon pilgrimage to Mecca are at least overt expressions of faith that reflects religious trust in Hausa society. In the open markets of Kano, trust in God is repeatedly shown by local traders. During daily prayer hours, shops in the markets are flung open and left unattended by their owners. Merchant trust in God is subsumed in the notion of His unfailing protection of business property, and the feeling that God-fearing customers will not loot vacant shops.

Second, trust in the primordial and ethnic settings is a cultural product derived from the failure of the Nigerian state and its corrupt apparatuses. Given the failure of the state to provide support or to protect individuals in the experiences that characterise daily life, ethnic members are crucial in providing emotional and other forms of support during times of distress. Trust, which is as much a habit as a rational calculation, has a large and measurable economic value (Fukuyama, 1996, p. 10). Third, the establishment of a predictable legal system provides a basis for trust. Invariably trust, predictability, and reciprocity[117] go hand in hand.

A fourth source of trust is derived from competence. In the factories, entrepreneurial trust in management and employees is based on their skills and competence in the handling of delegated tasks. Thus, intra-organisational relationships rest on technical know-how and the ability to complete delegated assignments reliably (Sitkin, and Roth, 1993). Couched in Luhmann's (1979, p. 41) terms, subordinates who want to show themselves as trustworthy must strive to prove the utmost industriousness, and readiness to carry out tasks loyally beyond the normal practice. Conversely, the inability to display loyalty or to approach an assignment in an acceptable way may lead to a distrust in the relationship between the worker and his/her immediate superior.

Distrust, which displays repertoires of fixed roles, entails seeing others as incompetent, unreliable, or deceitful.[118] The presence of incongruity between parties in a specific relationship can lead to distrust. However, 'trust is easier to transform into distrust than is distrust into trust'.[119] Once distrust is created about a particular system or person, a change of impression becomes difficult. Deviant behaviour such as theft and drug addiction often creates a long-lasting stigma[120] that inhibits trust. In view of this, people with previous crime records are often rejected as business partners or employees in many Nigerian firms.

Hope, however, occurs when the probability outcome is mostly unknown. On the other hand, trust carries a stronger connotation than hope. Fundamentally, trust deals with confidence and reliability, which hope does not necessarily entail (Giddens, 1995, p. 29). Trust, which is bound up with contingency, is a continuous state. Entrepreneurial trust is embedded in 'ontological security', a concept which refers to societal confidence in 'self-identity, and the constancy surrounding social and material environments of action' (ibid.). A critical analysis of cross-cultural variations in trust further reveals that 'trust is not anachronistic to the modern society but the *sine qua non* for its success' (Fukuyama, 1996).

Several reasons account for the presence of trust in a specific relationship or system. Trust is usually present because of mutual benefits to actors in a particular situation. Mutual trust is based on the norm of reciprocity and the feeling that others will conform to rules or behave according to expectations. Predictability is another reason for trust. Entrepreneurial perceptions that economic indicators in a specific environment will endure can be a good reason for building healthy trust relations. Thus, where there is predictability, there is a strong basis for trust among business groups. But this is based on the assumption that a particular system is dependable or reliable. From a cognitive perspective, trust as a mental representation may change after a series of interpretations of consequences and behavioural adjustments. From a moralistic perspective, the potentials for trust relations are strong where values are shared by parties.[121]

Basic Conditions for Trust

Certain preconditions are crucial for the endurance of the presence of trust (Huemer 1998, p. 98). Trust is possible when trusting expectations make a difference to a decision. Usually, in trust situations, there are critical alternatives in which a breach of agreement can result in harm. One who trusts, takes 'cognisance of the possibility of excessive harm' (Luhmann, 1979, p. 24). Furthermore, trust is possible 'where the potential damage' is greater than the advantage which is needed. A trust-related action may be taken where risk is within tolerable limits (Huemer, 1998, p. 100).

Familiarity is a precondition for both trust and distrust (Dasgupta, 1988). For business persons or groups to trust each other, they need some established form of social interaction and familiarity with each other. Conversely, familiarity with bad business records or untrustworthy behaviour of business colleagues, may heighten distrust. Familiarity of the trustee is a vital factor in economic transaction. Thus, trust often develops among groups after a long period of interaction. Luhmann further explains that an orientation with the past can 'simplify the world and render it harmless'. Accordingly, trust is only possible in a familiar world. Thus, the plausible assumption that the trustworthy person retains past record is present in nearly all trust relations. Human encounters are represented in history, because history and the past serve to reduce complexity. The 'reduction of complexity through trust becomes significant in an increasingly organised social structure' (Luhmann 1979, p. 25). Organised social structures provide both internally and externally guaranteed security, which serve as foundations for trust. History and reliable backgrounds are often relevant conditions for trust in business transactions. The reason is that familiarity and trust complement each other. Reputation for honesty and reliability in persons and systems accumulates over a long period (op.cit., 1988). The knowledge of a person's disposition is critical in trusting relationship. But this knowledge will only develop gradually with time.

There are both anticipated and unanticipated results and consequences associated with trust. A trust relation between truster and trustee is purposive (Coleman, 1990). In a trust relation, there is usually awareness of an alternative following the course of a particular action. This awareness about alternatives provides options for choice. Such situations often allow the exercise of precautionary and selective steps. There is no consensus on the relationship between trust and risk, uncertainty and vulnerability (Huemer, 1998, p. 103). Predictability is a basic condition for trust in a society. Conversely, insecurity and uncertainty serve to reduce trust in a given system. Although insecurity can sometimes strengthen contractual bonds and cohesiveness among business groups or coethnic members faced with common problems. For Deutsch (1958) predictability includes repetitiveness. In the business sphere, the belief that a system or person will behave in a specific way facilitates forecasting and planning. Trust relationships are usually enhanced by the consideration that 'we shall meet again' (Luhmann, 1979). Certainty and predictability reduce the possibility of a breach of trust relationship.

Arbitrary changes in macroeconomic policies by the government can lead to uncertainties that serve to reduce trust. Business associates can be trusted where there is adequate information and knowledge about their past activities. Without factual and background knowledge, trust relation becomes muddled. However, predictability is only a necessary, but not a sufficient, condition in trust relationships (Young, 1992). Any meaningful trust must transcend beyond predictability. Some antecedents for trust in society include predictability, motive, expectation, ability, and reputation.

Positive, not negative reputation and the evidence of enduring partnership are also very important in trust matters. Positive reputation has both economic and social value (Fama, 1980; Dasgupta, 1988). Usually exchange relationships acknowledge these factors. Reputation is thus, seen as a contract enforcer (Milgrom, and Roberts, 1992). In addition, it provides clues from which business partners can evaluate each other. Business reputation is an important tool for expansion. Thus, businesses with good reputation easily expand in their economic activities. In the absence of this, partners may resort to hunches in their evaluation process. Reputation is important when parties lack familiarity with each other (Madhock, 1995). However, reputation may be lost when parties renege on promises or fail to conform to rules. Potential sanctions, which include publicity of a bad reputation or a loss of the confidence necessary for daily transactions, may lead to deterrence. Societies are often characterised by the presence of low or weak forms of trust (Barney, and Hansen, 1994).

In Kano commercial networks, like elsewhere in the developing world, reputation is very important in ensuring the growth and sustenance of business. That certain family names have endured through the decades is perhaps a reflection of the reputation they have built over the years. Indigenous cola-nut merchants have survived in the Agadasawa ward of Kano because of their past reputation of honesty and trustworthiness. The honesty of this group of cola-nut traders has created a myth that has been transformed into market competitiveness. Most cola buyers will prefer a business transaction with Agadasawa traders because of their good commercial reputation.

Luhmann (1979, p. 33) argues that the truster seeks clues about whether trust is justified or not. Such clues include the extent of economic success in a particular environment or country. Of particular importance is the extent of success of local or indigenous business groups in the economic regime. In the manufacturing sector, investors will want to know the products which will sell quickly and with less cost in a specific market. In the factories, entrepreneurs will want to know the records of their employees, which groups to delegate responsibility to or entrust cash with. Employee honesty, competence,[122] hard work, and dedication to duty and punctuality may be the bases for which important roles are delegated to them. Clues, according to Luhmann (ibid.), act as safeguards to risk reduction.

However, exposure to a trust relationship often leads to vulnerability. Paradoxically, trust is based on a tacit notion that vulnerability would not be exploited (Sabel, 1993, p. 1133). Persons who trust in others also count on their goodwill. This goodwill hinges on a notion that there is no exploitative intent on the part of business associates. Although trust depends on predictability, dependability and faith, it is also a sort of 'gamble' or 'risky investment' (op.cit., 1979, p. 24). In spite of the many risks involved in trusting relationships, trust has no substitute in the daily conduct of human activity.

The Relevance of Trust in Economic Transaction

Modernity opens new opportunities which require both high risk and trust relationships (Giddens, 1995, p. 7). In particular, it creates more trust in abstract and expert systems. Trust in abstract systems takes the form of faceless commitments. This implies that 'blind faith' is sometimes necessary in short-term economic transaction (Kreps, 1990). Today, the use of travellers' cheques and credit cards in business transactions has increasingly led to entrepreneurial reliance on abstract systems on a global scale. New forms of transactions in the economic sphere have helped to reduce the complexity of carrying large sums of money across the world. Adaptation to abstract systems depends on trust. However, trust has to be learned (Luhmann, 1979). Learning to trust is a matter of gradually adapting to the changing circumstances of the environment.

Entrepreneurial low trust in abstract systems is often a reflection of uncertainty and ignorance. Inadequacy of information and lack of exposure to new methods can be a good reason for suspicion. Lewis and Weigert 1985) observe that there can be no reason for trust where there is absolute ignorance of a particular person, activity, or system. Given the low level of trust in abstract systems, transaction cost in the Nigerian society is sometimes increased by the desire for face to face transaction. Sometimes, the availability of bank cheques, and telephone facilities are not enough to deter the desire for face to face transanction among business groups in the Nigeria. This has often led to the vulnerability of local business persons to motor accidents and highway robbery.

In contrast, businesses in the industrially advanced parts of the world, have become less vulnerable to losses resulting from delays imposed by the need for face to face transaction. In addition, their economic activities have advanced beyond the weariness of theft and violent attacks by armed robbers. In a high trust society such as Sweden, banks do not require police guards even after working hours. This contrasts sharply with the situation in low trust societies of developing countries. As such, the banks in Kano and many parts of Nigeria are guarded around the clock to ensure the safety of money and other valuable deposits. In the urban business centres, including Kano and Lagos, it has become daily routine for armed police convoys to accompany cash to and from banks to ensure the safety of money and bank officials. The sight of police convoys accompanying cash in urban centres provides supporting evidence of the low level of trust in the Nigerian economy. In recent times, police convoys constitute a menace not only by obstructing traffic; they also often paralyse economic activities during much of the day in the cities. In the factories of northern Nigeria, police convoys are used on a daily basis to transfer cash to the banks for deposition and safe keeping. These activities encroach tremendously into business time necessary for productive activity.

Yet, trust is relevant because it reduces complexity and simplifies economic transaction. Suspicion and complications often set in where trust is absent. For instance, robbery attacks on traders and travellers in parts of Nigeria could probably be minimised or avoided if business transactions become less dependent on face to face transactions. A major handicap is that many entrepreneurs are not educated enough to trust the operation of abstract monetary systems and modern financial institutions. The preponderance of swindlers, con men and agents of advanced fee fraud (AFF) or 419 in the Nigerian business scene, continues to inhibit public confidence in abstract systems such as credit cards and bank cheques. Nevertheless, credit cards are already in use in a few Nigerian banks. It is pertinent to mention that such transactions are currently limited to trusted elite and credible business groups.

Despite the limited merits of keeping money at home, such attitudes have led to untold hardships and negative consequences for many local traders in the commercial sector. Perennial fire infernos in the markets of Kano over the years have robbed many traders of several million naira stashed in bags and metal trunks. Keeping money outside banks prevents the circulation of money necessary to stimulate production and growth in the economy. Today, indigenous entrepreneurs in Kano and parts of northern Nigeria are increasingly changing their attitudes and learning to trust and patronise abstract systems in their daily business transactions. This development has resulted in a rapid upsurge in the number of modern financial institutions.

In every society, trust is often the reason for the provision of credit facility to business partners. Every credit provision is based on the notion that partners will carry out fiduciary obligations without breaking or reneging on old promises. In its absence, credit facility may be narrowed or completely withheld. Investors' fears about customers reneging on their promises or exploiting their vulnerability may lead to the introduction of safety strategies and precautionary measures. A common strategy often applied by local entrepreneurs is the provision of narrow credit facility to customers as safety net. Sometimes little credit facilities may be provided by entrepreneurs as a way of ensuring personal credibility while preventing major losses from unreliable partners.

Business operators may withhold credit facilities from customers who cannot be trusted. In many developing societies including Kano, it is common to find inscriptions on shops or business outlets warning customers that there are 'no credit provisions here'. Sometimes building property or farmland could be pledged or mortgaged as collateral or medium of trust against credit facilities. Reneging on promises and cheating can militate against trust in business relations. Luhmann observes that trust educates. A business person who is cheated through trust will be cautious in entering a new trusting relationship. Economic opportunities and Business relations can be jeopardised without trust. Kreps (1990, p. 116) asserts that the trusted party will honour trust because to abuse it will preclude or limit opportunities to engage in future valuable transactions.

Observation of the Kano business sector reveals that, despite the prevalence of credit rotating schemes, and a tradition of advancing goods to brokers before financial settlements, there is low level of trust in the business environment. For many years, the spectre of mistrust has haunted Hausa and other Nigerian entrepreneurs. Interpersonal mistrust outside primordial circles accounts for the low level of joint stock and partnership ventures involving the extensive capital essential for establishing and developing large scale modern corporations. This low level of trust has inhibited the expansion of many businesses beyond the level of family control.

Trust is relevant for smooth day to day economic transactions. In addition, it encourages business expansion beyond family and geographical frontiers. In the contemporary era, trust opens access to new information about economic opportunities. Lack of access to adequate information, and opportunities for expansion inhibits the building of trust. Fukuyama's (1996) study shows that high-trust societies such as the United States, Germany, and Japan spawn large firms not based on kinship relations, whereas low-trust societies such as France, China, Italy, and South Korea typically operate smaller family firms. This shows a link between trust and the size of business firms.

The pervasiveness of patrilineal control of micro-businesses is not entirely due to inadequacy of capital but is also a reflection of entrepreneurial safety strategy. The desire for liquidity control among small-scale family indigenous businesses is derived from the pervasive distrust of the Western-style educated persons, or *Yan boko*. Educational difference between employers of labour and their subordinates can sometimes be a reason for distrust in the Nigerian business environment. Affluent but uneducated entrepreneurs sometimes exercise fears about well-educated assistants and managers exploiting their ignorance and vulnerability. The fear of 'educated outsiders' threatening control in family businesses has led to tremendous investments in the education of family members. Nowadays, many Hausa merchants and business industrialists boast and rely on well-educated family members for the conduct of their businesses.

Fukuyama's study shows that cultural diversities account for the differences in trust patterns across societies. Societies in Southeast Asia including Japan, he noted, are very suspicious of foreigners when it comes to economic matters. By contrast, the results of this study suggest some degree of inverse trust relationship among Hausa and Nigerian business groups. The inverse trust is subsumed in the notion of 'foreign is better'. Foreigners, rather than fellow-countrymen, are trusted and accepted in joint economic ventures. This peculiar characteristic does not particularly favour the expansion of indigenous businesses.

The extension of trust to foreigners in partnership ventures is borne out of several considerations. Foreign business partners are often perceived as having stronger financial wherewithal and easy access to venture capital. In addition, they are perceived as possessing vital technical expertise and information about trends in the international market that is crucial for business growth. Furthermore, given the respect among the foreign business community for indigenous laws, the propensity of reneging on initial agreements will be low. Nevertheless, the weakness of judicial institutions combined with the inability of the judiciary to enforce property rights and contract laws, further encourages business groups to revoke business agreements. Group formation does not depend on ethical habits but on establishment and proper functioning of legal institutions like property rights and contract law (Fukuyama, 1996). Efficient economic institutions are the cornerstone of development. Such institutions are devoid of arbitrariness, excessive nepotism, dishonesty, and mistrust.

Manufacturing activity, which requires large amounts of fixed capital, requires trustworthy partners and secured environment. An unsecured business environment can be a deterrent for investment in the manufacturing sector. Modern manufacturing also requires a great deal of division of labour. Very often traditional merchants, who exhibit distrust in the delegation of responsibility to subordinates, are less likely to invest in manufacturing ventures regardless of their financial positions in the society. Entrepreneurial interpretation of political and economic risk signals leads to diversification into multiple investments in the commercial sector. Distrust in the political economy is among the factors contributing to low investment in manufacturing ventures, which require long-term political and economic stability.

In the domestic market, low public confidence in locally manufactured commodities and consumer suspicion about the circulation of substandard and adulterated commodities are among the reasons for the low patronage of homemade products. The low degree of trust in made in Nigeria products sometime has a negative influence on local business expansion. On the international level, the current distrust in the Nigerian political economy has reduced the inflow of foreign investment and contributed to economic stagnation.

Trust in the business sector can be inhibited through suspicion. One method of evaluating the scale or level of trust, is the extent of joint or partnership ventures in the business environment. Such partners should not necessarily be family members or kin groups. The absence of joint ventures by non-family members in the Hausa society is explained by low level of trust. Sometimes unfounded fears about non-family members usurping control or being cheated by copartners may prevent the exploitation of certain opportunities. The level of trust exhibited by the Chinese entrepreneurs in the joint Chinese plastic venture covered by this study was very high. This trust is illustrated by the subsequent joint investments in hotel, steel, textile, industrial chemical imports, and sweet manufacture.

In patriarchal societies, where inheritance is partible, the fear of the family of diseased partners threatening separation and calling for the sharing of assets sometimes serves as inhibiting factors against trust and the setting up of joint businesses. Entrepreneurial weariness that business agreements even if documented will not be honoured has negated against the expansion of Nigerian and indeed African businesses. The absence of joint ventures accounts for the low injection of external capital relevant for the growth and expansion of indigenous businesses. Financial solvency is often the reason to trust in a business person. Financially bankrupt business persons are seldom trusted, especially with money or credit facility. Thus, the relevance of trust in economic transactions cannot be overemphasised.

Trust Via Clientelistic Networks

A fundamental way of developing business credibility is via the network of influential contacts (Macmillan, 1983, p. 8). A business network is a complex pattern of relations linking a defined set of persons, objects, and events (Dahl,

1998, p. 77). Network relations are characterised by the following: there is usually an informal access or corporation with a minimum amount of bureaucracy, actors work towards a common goal and mutual benefits, trust and personal relations are basic ingredients of these relations. In addition, there is often a common language that provides a basis for the smooth flow of information and interaction of members (ibid.).

Network issues loom large in economic transactions, not only in developing countries, but to the industrially advanced parts of the world as well. These culturally peculiar patterns of networks can be found across the world. In Japan, business networks are arranged vertically, between the large automobile enterprises and subcontracting units, whereas in parts of Europe, especially 'northern Italy, southern Germany and Denmark', the arrangements are horizontal (Dahl, 1998, p. 78). Essentially, some networks are developed for handling turbulent market situations while others for production (ibid.). Network patterns in Kano and parts of northern Nigeria operate on several levels and therefore are complex to define. By implication, network patterns often assume specific characteristics of the prevailing situation. As such, the network relations can described as hierarchical, opportunistic, and commercially oriented.

Entrepreneurs use trusted contacts to increase their own credibility and trustworthiness. Luhmann (1979, p. 64) argues that trust accumulates a kind of capital which opens more opportunities for more extensive action. In times of raw material scarcity and acute oil shortages, entrepreneurial networks in the plastic industry are important in providing information on markets where raw materials or illegal petroleum dealers can be found. Through a network of contacts, plastic manufacturers in Kano can purchase recycled plastic materials from the informal markets of Chad, Cameroon, and the Niger republic. In addition, network relations are crucial in providing market information about prices and the performance of competitors.

Religious networks provide avenues for establishing trust or confidence with trading partners (Paden, 1973, p. 12). Sometimes, it is a medium for introducing clients to new economic opportunities. In the Hausa community, personal piety and attachment to faith are important in providing a kind of security and access to credit facility.

Religion is relevant because of the notion that a God-fearing person seldom reneges on his/her promises. Religion enhances interurban marriages. Familial background, class, and economic success often provide conditions for such marriages. Marriage and familial fraternity among successful entrepreneurs, aristocrats and ruling classes in the economy of northern Nigeria are well documented by Tahir (1975) and Paden (op.cit.). Marriage ties in Kano often transform into economic bonds. In the business sphere, persons could provide fronting roles for the ruling classes. Furthermore, marriage ties with ruling and local *malam* classes may be exploited to boost personnel prestige, reputation, and trust in the urban economy.[123] The marriage patterns, which ensure the exchange of one elite group with another, are consonant with Pareto's notion of 'circulation of elite'. Clientelistic networks and social contacts are avenues for gaining employment in the modern industrial sector or access to the market.

Among traders and business industrialists in northern Nigeria networks and alliances with ruling classes, politicians and religious clerics are the instruments for enhancing trust.[124] Social networks can become a means through which personal position can be enhanced or consolidated. Sometimes business groups are compelled to identify themselves along political, quasi-political associations or with high ranking military classes to enjoy special favours. Special favours often include access to favourable foreign exchange rates and state contracts. In order to understand the operation of network relations, it is necessary to examine how foreign investors react to the socio-cultural milieu.

Client-patron relationships and networks of alliances in northern Nigeria are not restricted to local business groups. Expatriate investors particularly require network relations to secure business partners and relevant information about the markets in which they operate. Olzak (1983) posits that the process of entry into new society by foreign business investors puts them in distinct positions often in conflict with the interests of local business groups which often hasten ethnic group formation. This assertion is particularly valid of Lebanese, Syrian, Indian and Chinese business groups in Kano and parts of Nigeria. Research on immigrant businesses further suggest that, the more deeply embedded expatriates or foreign business groups are within kin and coethnic networks, 'the more salient ethnic group boundaries will be' (Waldinger, *et al.* 1990, p. 36).

Like other economic actors, expatriate business operators do require influential contacts with local powerful groups to ensure hitch-free economic transactions crucial for business success. Informal contacts with ruling classes, law enforcement agencies, and tax officials help to strengthen economic transactions. Lebanese investors are in the forefront of many investors when it comes to alliances with powerful ruling and political classes. The Lebanese business community in Kano is often well connected with the traditional ruling classes.

Among indigenous business groups, personal trustworthiness may be enhanced through intermediaries and mediators. This implies that sometimes trust can even be bought. When entrepreneurs patronise traditional or chieftaincy titles in Kano and other parts of Nigeria, they do so with a purpose and clear mission. Indigenous entrepreneurs sometimes invest heavily in traditional titles in order to enhance personal ratings and trust in the local community and sometimes across the nation. A traditional title is a social capital that can transform into an economic capital. The evidence in northern Nigeria suggests that a traditional title may provide access to credit facilities in the banking sector. It may also create new economic networks with multinationals, alliance with government functionaries, and the winning of federal, state, or local government contracts. A person's image is therefore an important instrument for trust.[125] This is because the social expectations placed on the traditional titleholder expects that such an individual must be a trustworthy person who can be entrusted with public property and key contracts. Nevertheless, the Nigerian business scene shows that sometimes such jejune assumptions have led to the misappropriation of public funds with great disappointments. This implies that sometimes trust can be won through deceitful self-presentation.[126]

Closely tied social networks are crucial in providing the confidentiality for credit-raising mechanism or access to venture capital (Light 1972). Small brokers, or *dillalai*, and petty traders without influential contacts in search of credit or venture capital often rely on locally trusted persons as intermediaries. The presence of a reputable person in a specific business negotiation may be instrumental in gaining access to goods and services before financial settlements are made. Clientelism, networks, and access to influential contacts appear relevant for establishing trust in the urban Kano business sector. In the economic arena, indigenous cultures sometimes inhibit development of trust by failing to see the expansion of business beyond family frontiers.

Nevertheless, trust is learned. This learning process, as Luhmann (1979) observes, is continuous because new situations and new people are continually posing problems of trust throughout life. As we enter the next millennium, new situations will influence the development of new attitudes regarding business trust. The preponderance of businesses and vibrancy of micro-economic enterprises suggests a presence of a modicum of trust. This is because business transactions or economic investment is possible only where there is some amount of trust in the future. In the commercial sector, the cultural form of advancing goods to brokers or *dillalai* before financial settlements also confirms the presence of trust. However, the pattern and level of trust as observed in Kano and in the Nigerian business sector is generally low.

This low trust is a product of several factors including, unpredictable political climate, inconsistencies in macroeconomic policies involving exchange rates, currency allocation, industrial relations, and business fraud. These combinations of factors interact negatively to discourage the building of a strong trust necessary for the growth of the business sector. For trust to be present in a system or activity, an amount of certainty is needed on how future events will turn out to be. Given the uncertain situation in many low trust societies including Kano, the family business remains dominant. Entrepreneurial reluctance to subordinate personal interest by involving external professionals is borne out of these uncertain and unpredictable considerations.[127]

Summary and Conclusion of Part One

In Chapter One, I provided an outline and mainstream understanding of the study, with a theoretical background on culture's relevance to economic performance. This was done through a reexamination of various cross-cultural paradigms and their relevance to investment patterns across the world. In the introductory Chapter, the method and unit of analyses deployed in the study were also discussed. The discussion in Chapter Two, however, was on entrepreneurship and its essential attributes. In the same Chapter, the historical evolution of indigenous entrepreneurship, precapitalist, and modern forms of venture capital were examined by attempting to bring analytical sociological perspectives to bear. Entrepreneurial risk-reduction attitudes as observed in the Chapter, were reflected in multiple investment strategies. It was observed that the operation of multiple businesses by entrepreneurs sometimes inhibited the concentration of capital necessary for investments in large-scale economic activities.

In Chapter Three, an attempt was made at integrating the subject of culture outlined in Chapter One, with entrepreneurial rationality. The basic reason for this was to provide a theoretical underpinning for empirical analysis in the Chapters which will follow. The framework of analysis in Chapter Three was focused on the logic between religious reform and new forms of entrepreneurial rationalisation in northern Nigeria. It was observed that not only is Islamic puritanism receptive to modernity, it also positively encourages entrepreneurial pursuits in commerce and modern industrial activities in Kano and elsewhere in the Islamic societies. In Chapter Four, the study looked more closely at entrepreneurial trust relations. Key conditions and relevance of trust in business transaction were examined in this Chapter. Although the general trend in trust patterns was low among business groups, the reasons for this were not entirely cultural. Other reasons identified for the low level of trust include pervasive uncertainty in the political, economic, and social life of Nigeria. In the face of uncertainties, entrepreneurial networks and clientelistic relations were seen as crucial in providing information on new market niches, sources of raw materials and scarce petroleum products in the manufacturing sector. Networks also provided informal avenues for employee recruitment in many firms.

Part II
The Nigerian Plastic Industry, Entrepreneurial Profit Motive, and Resource Controls

5 The Nigerian Economic Environment

The Nigerian economy during the last quarter of this decade has been anything but impressive. The discrepancy in the political and socioeconomic climate in which this research took place suggests that most businesses were passing through critical stages of their history. Several local and external factors were responsible for the murky political and economic environment. It is worth mentioning however, that the local factors were very overwhelming. On the local front, the military annulment of the June 1993 elections, which would have restored democracy, plunged the country into turmoil. The prolonged political impasse, which brought the economic life of the country to a near-standstill, threatened to make Nigeria an international pariah. Available data from the Manufacturers' Association of Nigeria (MAN), shows that Nigeria's worst economic performance in nine years was recorded in 1994.

All sectors of the economy during this period were at a standstill or nearly paralysed.[128] Not surprisingly, the manufacturing sector recorded a 5 percent decline during this period. The unstable macroeconomic environment, a deteriorating infrastructure, and the scarcity of foreign exchange led to a decline in capacity utilisation in the industrial sector. As will be expected, the general decline in economic activities, resulted in redundancies and retrenchments in many factories including the three plastic firms that form the basis of our discussion in this Chapter.

Despite the vaunted projections of the Federal Government, there was a substantial budget deficit during the period. As of June 1994, the Central Bank of Nigeria (CBN) had reported a budget deficit of N30.227 billion, equivalent to 56 percent of the total federal revenue. High interest rates in the banking sector and the significant depreciation of the naira in the parallel market ignited a sharp increase in the rate of inflation.[129]

Table 5.1 shows that consumer price inflation rose from 7.4 percent in 1990 to 57.0 percent in 1995. Figures from unofficial sources show that the inflationary rate during this period was in most cases over 100 percent. According to a World Bank report, the high inflationary trend was due to the govern-

Table 5.1. Nigeria's Economic Indicators 1990–1994

1990	1991	1992	1993	1994	
GDP at market price N bn	260.6	324.0	549.8	697.1	897.5
Real GDP growth %	8.2	4.7	3.0	2.3	1.3
Consumer price inflation %	7.4	13.0	44.6	57.2	57.0
Population m	86.7	88.5	91.3	93.3	95.2
Exports fob $ bn	13.59	12.25	11.79	10.90[b]	10.65[b]
Imports fob $ bn	4.93	7.81	7.18	6.80[b]	6.60[b]
Current account $ bn	4.99	1.20	2.27	·1.81[b]	0.75[b]
Reserves excl gold $ m	3,864	4,435	967	1,372	1,386
Total external debt $ bn	34.54	34.44	31.00	32.53	n/a
External debt-service ratio %	23.6	26.6	29.4	n/a	n/a
Crude oil Production m b/d	1.84	1.95	1.88	1.91	1.90
Manufacturing output index 1990=100	100.0	109.3	111.0	9.7	94.4[d]
Average exchange rate (av) N:$	8.04	9.91	17.30	22.07	22.00

Source: The Economist Intelligence Unit, London, July, 1995.

ment's inability to control 'fiscal deficit and its recourse to excessive monetary financing'.[130] The high interest rates in banks were also unsuitable for business initiative and profitability.

Compelled by unprecedented interest rates in 1993, the government introduced budgetary measures making it mandatory for banks to peg the savings/deposit rates at 12–15 percent per annum. Although the maximum official lending rate was pegged at 21 percent, in reality the rate was between 40 and 60 percent (CBN, 1994). Financial malpractice in the banking sector compelled the government to close down some commercial and merchant banks during this period. The low level of business confidence in the economy meant that there was more money in the hands of the public than in the banks, a trend that continued to fuel inflation in the economy. The inflationary situation in Nigeria during this period led to a vertiginous fall in the value of the naira against major foreign currencies such as the American dollar and the British pound (ibid.).

In retrospect, the world recession, with its negative impact on the economic growth of many countries, was witnessing a rebound during the period under review. The United States of America recorded an annual growth rate of 3.3 percent in its gross domestic product. Other countries that recorded growth during the same period included Germany, France, Canada, Italy, Australia, Britain, and Holland. A significant dimension in this trajectory of recovery was the case of some Latin American and Asiatic countries including Mexico, Chile, Argentina, China, and India. In sub-Saharan Africa, the story was different. Most of the countries in this zone 'including Nigeria, were left behind in the rising tide of growth and recovery'.[131]

Nigeria's natural resources include petroleum, tin, columbite, iron ore, limestone, lead, and zinc among others. Oil, which accounts for 93 percent of Nigeria's export earnings, and is the main source of government revenue, has become the dominant factor in the country's national life (Ikein, 1990). The oil sector provides half of the Gross Domestic Product (GDP) and nearly 80 percent of budgetary revenues. With the negative developments in the international oil markets, and mounting external debt, the Balance Of Payments (BOP) position grew worse. While the world economy continued to grow, the price of oil in the world market rather witnessed a decline. This was due to non-adherence to production quotas by the Organisation of Petroleum Producing Countries (OPEC) and non-OPEC members resulting in increasing global oil production. The world economic recovery, therefore, did not benefit the Nigerian economy. Consequently, foreign exchange inflow dropped from $3,223.0 million in January-May, 1993 to $2,566.4 million in 1994. This represented a 20.4 percent decline in foreign revenue. During the same period, the oil workers' strike and lax maintenance of machinery also reduced the production and supply of oil, leading to a registered drop of 0.39 million barrels per day.

Consequently, the decline in foreign revenue had a negative influence on the business sector, which relies heavily on imports of technical equipment and raw materials from the outside world. Significantly, the most appalling performance was recorded in the private non-oil-exporting sector. Total income of this sector showed a decline from $689.8 million in January to May 1993, to $395.8 million in 1994 (ibid.). The dynamism associated with the Nigerian business sector ebbed during this period, culminating in a great anxiety among investors. Political risk signals that continued to influence business decision-making during this period led to entrepreneurial ultra-precautions and aversion for new investments. The scenario also led to a flight of capital to economically and politically safer parts of the world. The conclusion to be drawn from the distressed economic and political climate is that all spheres of the economy, and particularly the business sector, experienced a decline. Against this difficult socioeconomic and political background, the study of the three plastic firms took place.

The Plastic Manufacturing Industry

Plastic manufacturing is a branch of technology that deals with the use of rubber, synthetic resins, polyvinylchloride (PVC) compounds, and petrochemical materials for the manufacture of a variety of products and objects according to the desires of the client. The plastic industry in Nigeria belongs to the National Union of Footwear, Leather, and Rubber Products. A variety of items produced in the plastic industry include household materials such as water heaters, tanks, mattresses, sleeping mats, cushion seats, buckets, cups, plates, spoons, pet bottles, cosmetic containers, rubber bags, footwear, and toys. Other products of the plastics industry include radio and television components, film, cassettes, extrusion pipes, car tyres, bumpers, and steering wheels. Official records from the Kano State Ministry of Industry show that in 1995, there were over 60 firms engaged in plastic manufacturing activities within the Kano metropolis. Of this figure, Levantine and Asian investors owned approximately 67 percent.

The three case studies featured in this Chapter are based on entrepreneurial ethics and trust in the plastic-manufacturing sector. For the purposes of the study, the three plastic firms have been classified as A, B, and C. As indicated earlier, these include, indigenous Hausa, Lebanese, and Chinese plastic firms. The alphabetical designations and entrepreneurial backgrounds, rather than pseudonyms, are used interchangeably throughout this study. In view of this, the actual names of the firms will remain anonymous in conformity with the requests made by their owners during my fieldwork.

The first industrial layout in Kano was created in 1955 by the colonial administration in an attempt to promote local processing of agricultural and export commodities. Since then, different administrations in the post-independence era have developed four additional industrial locations to cope with the ever-expanding industrial sector. Currently, manufacturing and processing activities occur in five major locations in Kano. These include Fagge/Gwammaja, Sharada, Bompai, Hadejia-Road, and Challawa industrial estates. These industrial locations are widely dispersed. Challawa is the farthest from the centre of the city, at twenty-six kilometres. The process of selecting the case studies ensured that the manufacturing firms came from different geographical locations.

Scenically, the Bompai area, which is noisy and overcrowded presents a remarkable contrast to the more tranquil Challawa and Sharada industrial areas. The Bompai industrial zone is notable for textiles and textile-allied manufacturing and processing activities. Many business enterprises, including insecticide and perfume manufacturing factories, are found in this area. The Chinese plastic factory examined in this study was selected from this vicinity of predominantly textile manufacturing activity. The growth of factories in the Bompai estate over the years has given rise to the development of an industrial settlement at Tudun Wada. Many of the early settlers in this area were industrial workers or wage earners and their families. Today, Tudun Wada has grown to accommodate all categories of people including petty traders and non-wage earning peasants.

Sharada industrial estate lies about two kilometres outside the old walled-city of Kano. The industrial area, which was created by the Kano State Government in the early 1970s, is divided into three 'phases', most of which are occupied with industrial buildings and economic activities. Because of the diversity of manufacturing and processing activities, the factories at Sharada range from soft drink bottling, the manufacture of sweets and candies, sugar, cosmetics, soaps, dry-cell batteries, textiles, wire, nails, galvanised iron sheets, motor parts, metal containers, fuel tanks, mosquito coils, and plastics. The Hausa plastic firm, which is one of the case studies in this study, was selected from the 'phase one' of the Sharada industrial area. There is now an appreciable increase in the number of agro-allied, food processing, and pharmaceutical firms along the Hadejia Road.

The second case study, the Lebanese plastic factory, was selected from the Challawa industrial area. Like Sharada, the Challawa industrial zone was created by the Kano State Government to cope with the expanding manufacturing sector, reduce congestion, and limit the disposal of chemicals in the centre of the city. Unlike the Bompai and Sharada areas, which are sandwiched between private estates, the Challawa zone is contiguous in terms of industrial buildings. The most predominant activity in Challawa is leather tanning, although there are meat, paper processing, soft drink, ceramics, and plastic manufacturing activities. Only two-thirds of the firms in these industrial locations were operating at the time the fieldwork was conducted. Of this figure, nearly one third of them were producing at 30 to 45 percent of their capacity due the scarcity of raw materials and spare parts. The reason for this was the economic recession, and the accompanying foreign exchange scarcities, a scenario that further led to employee redundancies and an unprecedented rise in inflation in the Nigerian economy.

Although Nigeria is a major producer of oil and has abundant reserves of petrochemical products, local initiatives towards production of plastic raw materials have not been encouraging.[132] The establishment of a petrochemical factory at Warri under the Nigerian National Petroleum Corporation (NNPC) has not resulted in any significant improvement in the local manufacture and supply of plastic raw materials. Approximately 79.4 percent of the raw materials used in the Nigerian plastic industry come from abroad, while the remaining 21 percent are obtained locally.[133] In the local market, the prohibitive prices of imported plastic raw materials showed inflationary tendencies. A ton of 'virgin' plastic raw material, purchased at N950 or approximately $12.00 in

1985, sold at N190,000 naira or approximately $2375 per ton in 1995. Apart from the food, beverage, tobacco, wood, nonmetallic mineral, and textile industries, many industries in the manufacturing sector showed heavy reliance on raw material imports from abroad.

The raw materials used in the Nigerian plastic industry are usually fresh synthetic polypropylene, simply known as 'virgin materials' from foreign countries such as Brazil, the United States of America, Britain, and the Ivory Coast. Virgin plastic materials are usually bought from European and Levantine dealers living in Nigeria. Locally, the plastic manufacturers rely on indigenous merchants and the garbage industry for the supply of recycled materials. The dealers of recycled materials merely serve as intermediaries between garbage collectors and the plastic firms. Garbage picking in recent times has become a full-time occupation for the urban impoverished and people exposed to the vicissitudes of the everyday struggle in Nigeria.[134] Most garbage collectors are 'child or teenage entrepreneurs'.[135] Some of these child entrepreneurs are local Islamic students, or *almajirai*, who can be found on a daily basis searching on garbage heaps, rubbish dumps, and fetid pools for metal and plastic scraps. The recycled materials, which are bought cheaply from garbage collectors, include discarded plastic buckets, shoes, sandals, crates, and extrusion pipes.

In the factories, the materials then undergo a cleaning process to be separated from dirt and foreign bodies before being ground into flakes by workers using grinding machines. Shop-floor workers in plastic factories use magnetic instruments to manually separate the rubber flakes from iron or metallic objects, which may hide in the plastic flakes. The plastic flakes, which are added to virgin materials, are then subjected to intense heat and melted by machines in the manufacturing process. The materials used in plastic manufacture are therefore heterogeneous mixtures of virgin and recycled materials. The mixing of cheaper recycled plastic materials with more expensive virgin components has become a cost-reduction strategy in the Nigerian plastic industry. Cost reduction is one major means by which manufacturers can sell their products at locally affordable prices. Nevertheless, the scarcities of raw materials and the astronomical increases in their prices are major constraints to manufacturers in the plastic industry. Entrepreneurial reliance on imported raw materials has meant that, in the face of dwindling revenues and foreign exchange scarcities, most factories are compelled to operate below their normal capacity.[136]

The plastic industry, for instance, showed a decline in capacity utilisation from 40.8 percent between July and December of 1993, to a mere 36.4 percent during the same period in 1994. Although there were differences in capacity utilisation between indigenous and expatriate firms with the situation favouring the latter, such data was not readily available.

Factory A: An Indigenous Plastic Manufacturing Firm

The Hausa plastic firm or 'factory A', began as a sales outlet of plastic household materials in 1966. Economic success of the commercial venture led to a diversification into manufacturing activity in 1982. The local plastic factory was completely managed and supervised by Nigerians. There was no external board of directors in this firm. By Nigerian investment provisions, the company law is applicable to all the transactions in this firm. Although the installation of machines and equipment started in 1980, the actual production of plastic materials began in 1982. The production process, which started with 50 employees and four machines, expanded to 120 employees with 16 machines in 1994. Out of the total number of employees in this factory, 34 were included in the sample of respondents.

A breakdown of types and functions of machines revealed that there were four blowing machines, nine injection machines, and three grinding machines. Initial capital investment in this firm in 1980 was less than one million naira. However, with the high rate of inflation in Nigeria, the capital investment in the firm stood at over 40 million naira in December 1994. A classification incorporating the current inflationary rate puts the firm in the category of medium-scale firms although, by actual assessment, it belongs to the category of small-scale firms. With one exception, the entire management staff of this firm was related to the factory owner. Most of the workers in this factory were Hausa and hailed from the same local government area as the entrepreneur.

The local factory specialises in the manufacture of household products such as water tanks, buckets, kettles, or *buta*, for Islamic rituals, traditional spoons, or *ludai*, cups, bowls, bottle-caps, cosmetics containers, and plastic nasal inhalers for a pharmaceutical company. The management of the indigenous Hausa factory could adapt or manipulate the imported technology to suit cultural needs. Some household plastic products were unique to the firm. Although the machines and technical equipment came from Italy and France, the iron moulds used in them were locally fabricated, according to the standards and tastes of the management of the factory. In particular, manufactured products such as the Hausa special spoon, or *ludai*, plastic bowl, or *kwano*, and large kettle, or *buta*, for Islamic rituals, follow local design standards. This positive development lends credence to the assertion that 'every manufactured product is the ambassador of a certain culture' (Gbotokuma, 1992).

Local ingenuity in the Hausa firm led to tremendous sales of the firm's products in the mid-1980s. However, the success of the firm attracted imitation from competitors, which consequently resulted in poor sales in subsequent years. Products of local plastic factories in Kano are mainly sold in domestic and regional markets. Due to the absence of sales and marketing departments in this firm, the local director and his accountant combined their many managerial and accounting functions with sales roles. The absence of division of labour in this firm made the execution of tasks haphazard.

The managerial pattern in the Hausa firm was patrilineal. This approach, which is quite rare in modern industrial management, is probably considered best suited for Nigerian workers. According to Fashoyin (1980, pp. 1–19), most Nigerian workers want recognition and when it is given, they concede to the seniority of their employer. In the patrilineal method, the entrepreneur assumes the role of the parent or benefactor of his employees. Most workers in the patrilineal system seldom follow the organisational hierarchy. Aggrieved workers in need of certain favours consult directly with the proprietor of the firm without having to pass through several tiers of hierarchy. This practice creates a congenial working environment and a feeling of acceptance among the workers. Although the strategy works well in commercially-oriented enterprises, its suitability for the modern manufacturing sector remains doubtful. A major problem associated with this practice is that sometimes the assumptions and measures taken for employees may not be in their own interest.

In 1984, the federal government imposed a freeze on wage increases in both public and private sectors of the economy. Despite the reversal of the wage freeze policy in 1990, which created hope for a new minimum wage, employee wages in both the public and private sectors of the economy remained unchanged. The problem was further compounded by the fact that, in Nigeria, wages do not usually keep pace with inflation and the general cost of living. The average monthly salary of employees in the plastic firms, which is N1,200.00 (approximately $13.00), is barely enough to sustain them for two weeks. Despite the incredibly low wages, the relationship between management and workers was cordial. This can be attributed to the strong kinship ties and the patrilineal style of management in the firm.

Factory B: A Lebanese Plastic Firm

The Lebanese plastic-manufacturing firm, like most Levantine businesses in Kano, is family-owned. Business records showed that the firm was established in March 1978. Construction of the factory began the same year and ended in 1979. Production in the firm commenced in 1980, two years ahead of the Hausa firm. By the provisions of the Nigerian Enterprises Promotion Decree (NEPD) of 1972 and 1977, the firm was expected to have a minimum Nigerian interest of 40 percent equity in joint participation. The law, however, made room for full ownership and control by expatriate owners who naturalised as Nigerians or Organisation of African Unity (OAU) citizens. Through legal loopholes and 'sloppy implementation', of the Indigenisation Decree, most Lebanese business owners in Kano acquired naturalisation and multiple OAU citizenship[137] in the 1970s to retain full ownership and control of their businesses without outside meddling. The Lebanese proprietor of the plastic factory was spared the ordeal of joint ownership by dint of his Nigerian citizenship.

Hoogvelt (1979, pp. 56–68) observes that apart from naturalisation, owners of expatriate firms had several ways of conveniently avoiding the indigenisation decree. One of their strategies was to use influential indigenous people as 'front men'. Expatriate investors, especially Lebanese business proprietors, who could not escape the decree sometimes personally selected their shareholders to ensure strong personal control over the affairs of their businesses (ibid.). In extreme reactions to the decree, some Lebanese investors sold their businesses and left the country, but such cases were few and isolated.

A common practice of foreign firms in Nigeria is to include indigenous persons on its board of directors to manage its affairs; this strategy was absent in the Lebanese firm. The Lebanese plastic company has subsidiaries in unrelated economic fields, such as industrial chemical imports, the export of skins and hides, and a textile commercial outlet operating at the Kantin kwari market in Kano. The business subsidiaries, which were run by other members of the family, operate independently of each other. It is necessary to mention that the plastic firm was the most prominent of all the business enterprises owned by the Lebanese entrepreneur.

The Lebanese plastic firm originally specialised in the manufacture of household materials such as rubber shoes and sleeping mats. Production of rubber shoes in this factory stopped due to a drastic fall in demand in the late 1980s when plastic shoes were no longer in vogue even with rural folks. Total capital investment of the firm was estimated at around 60 million naira (approximately $722,891) in 1994. As of June 1995, the firm manufactured only sleeping mats. The number of employees in the Lebanese firm stood at 247 in May 1994, out of which 57 were included in the sample of respondents. The raw materials used in this factory were similar to those used in the indigenous plastic factory. They include propylene resins, PVC compounds, thread yarns, and 'masterbatch concentrates'.

Unlike the Hausa plastic firm, the Lebanese plastic firm had direct access to raw materials from overseas countries such as the United Kingdom and Brazil. Direct purchasing of raw materials from overseas countries, without having to pass through intermediaries, made the cost of production in the Lebanese firm lower than in the Hausa firm. Furthermore, direct importation of raw materials also implied that the problem of scarcity of raw materials in the Lebanese firm was not of the same magnitude as in the local Hausa firm. Mat weaving entails the use of thread yarns to hold plastic fibres together. The factory purchased its thread yarns from a textile producing company at the Bompai industrial estate. The management of the Lebanese firms considered the production of 24 huge bales of woven mat per day satisfactory, especially given the scarcity of raw materials.

The Lebanese firm exerted stricter managerial control than the indigenous Hausa firm did. There were only three managerial positions, which were shared by the owner of the firm, his junior brother, and a Nigerian accountant. None of the supervisory and technical staff in this firm exercised any significant influence over other workers on policy matters. Managerial strategy was also non-bureaucratic, perhaps due to the small size of the firm. The salaries paid to workers were low, comparable to those paid in the Hausa plastic firm. Most of the line employees in this firm, like those employed in the Hausa firm, had very little education. Usually the level of education was never beyond secondary, technical and Quranic schools. Some tasks performed in the plastic factories were simple and therefore did not require high educational qualifications. In spite of the occasional grudges by a few workers on the shop floor about poor salaries, and working conditions, the working climate was quite cordial.

In Kano and other parts of Nigeria, Lebanese investors are considered shrewd businesspeople, guileless in their business dealings. Most Lebanese business operators in Kano are seen by many as tax defaulters. Their business transactions are therefore highly secretive. The Lebanese are also believed to possess dual citizenship and 'mysterious ways of surviving in business' (Beckman, 1982, p. 40). Their unconventional approach to business apart, the Lebanese of Kano and in other parts of Nigeria are generally successful in nearly all fields of endeavour. Lebanese prosperity is attributed to the ability to soften bureaucratic rigidity and bypass formal rules and regulations. Thus, the Lebanese business community in Kano possess boundless energy and a long commercial pedigree. Of all expatriate groups in Kano, the Lebanese are the most easily assimilated. The majority of Lebanese and Syrian investors who are second generation Nigerians, regard themselves as citizens of Kano and do not often show any strong desire to return home at least in the near term. Their ability to speak the local language, manipulate the law, while freely mixing into the local milieu, often makes them heroes rather than villains.[138]

The Lebanese business community in Nigeria is well connected with high government functionaries. Consequently, official malpractice involving the illegal lifting and export of crude oil to undisclosed destinations, are conducted through the assistance of Lebanese business people who often act as front men.

Unofficial sources from the Nigerian business circle suggest that for every one US dollar illegally earned by a government functionary through a shady oil deal, there is an unknown Lebanese partner who earns twice as much.

Although most Lebanese residents in Kano are professing Muslims, most of their economic activities are contradictory to the Islamic ethics. Despite the Islamic prohibition of gambling, Lebanese Muslims own nearly all the casinos, gambling pools, and betting houses in Kano. Male Lebanese textile traders in the Kano commercial sector often conduct their business activity with their wives and daughters. This is a business ethic that is rare in the Muslim Hausa society. It thus seems that, among the Lebanese resident in Kano, business interests tend to supersede religious considerations.

The monthly contributions made by manufacturers for the welfare of industrial workers, known as the National Provident Fund (NPF), were completely violated in the Lebanese plastic factory. To keep wages low and the workforce isolated, Lebanese business owners prohibit unions. There is also a higher rate of job insecurity in Lebanese than in public and multinational firms. Approximately 57 percent of the employees in the Lebanese firm were treated as casual workers. Among this category of workers were some employees who had served the firm for up to ten years but had no formal employment letters. Lebanese business operators usually feel that with employment letters, workers develop a feeling of security and start creating problems. Intransigent and vocal workers in Lebanese factories are denied employment letters as a deliberate strategy to weaken their radical stance. In view of this, only loyal employees are given employment letters in many Lebanese firms. For most factory workers however, employment letters are relevant because they make working relations legally enforceable. In addition, they signify recognition and offer emotional gratification.

Due to lack of proper records and bookkeeping practices, the remuneration system in the Lebanese firm was arbitrary and lacked any defined criteria. The salary of the individual worker largely depended on the extent of closeness with management. The lack of a well-defined system of remuneration became a source of discontent and antagonism among employees performing similar tasks in this firm. Nevertheless, most factory workers in Kano are more favourably disposed to employment in multinational, Nigerian or Lebanese firms than in Chinese firms.

Factory C: A Chinese Plastic Firm

The Chinese company is perhaps the most successful of the three plastic firms covered by this study. This plastic firm started its operations in 1977 as a joint business enterprise between two Chinese settlers in Kano. Government indigenisation policies of the 1970s compelled the Chinese management of this firm to divest 40 percent of its equity to indigenous partners. Although the company had an external board of directors that included indigenous and expatriate business people, the board exercised no significant influence in the affairs of the firm, apart from the occasional meetings it held during the year. The degree of involvement of indigenous shareholders was therefore minimal. By contrast, the involvement of Chinese family members in the control of the affairs of this firm was overwhelming. Fukuyama (1996, p. 56) explains that the essence of Chinese Confucianism is the family. As such, there is low degree of trust among the Chinese outside family circles. Hoogvelt (1979, p. 64) observes that the reason behind the strong influence of expatriate investors over their businesses is partly due to their selection procedures of share holders. She explains that shareholders are 'handpicked from such backgrounds as would minimise their future meddling'.

Unlike the Lebanese, the Chinese are not easily assimilated into the social life of Kano except in the factories or businesses. The reasons for the inability of the Chinese to assimilate may be cultural and religious. In spite of their inability to assimilate, the Chinese are respected for hard work and reverence for authority. Nevertheless, the widespread feeling that the Chinese overwork their subordinates discourage most Nigerian job seekers from taking up employment in Chinese factories.

With only 115 workers, in its initial operations, the Chinese firm has expanded to the point that it now has subsidiaries in Kano, Kaduna, Lagos, and Port Harcourt. With its subsidiaries in other parts of the country, the total number of employees in the Chinese company stood at over 3,000 in June 1995. The Kano branch of the plastic factory employed 453 workers out of which 75 employees were included in the sample of respondents.

The Chinese plastic factory in Kano produced indoor plastic footwear, commonly known as slippers. Due to the meagre incomes in northern Nigeria and other developing societies, many people cannot afford proper shoes. Therefore, the plastic slippers, meant for indoors, have become very popular, and are worn even in public places. For many years, the Chinese plastic factory, by far the biggest of the three firms, has enjoyed a monopoly over footwear production in Kano and northern Nigeria. However, the recent economic turmoil has affected the operations of the firm. Although the production standards in the Chinese firm were higher than the indigenous Hausa and Lebanese firms, some local constraints prevented the firm from achieving its full capacity level.

The raw materials used in the manufacture of plastic footwear in the Chinese factory were mainly 'virgin polypropylene resins', recycled footwear rubber materials, and 'masterbatch concentrates' which were obtained from both local and international sources. The procurement of 'virgin raw materials' directly from China reduced the cost of buying through secondhand dealers. At the shop floor, the plastic materials are first melted in the oven, made into rubber sheets, and left to harden for over a week. Shop-floor workers later cut the rubber sheets with special machines into various sizes. Because of the different processes involved in the manufacture of footwear, the factory was divided into sections, which included the production line, the mechanical section, and carpentry workshop. The carpentry section designed the wooden pallets on which molten rubber sheets were stored for other manufacturing activities.

Although management of the Chinese firm was strongly controlled by family members of the joint venture, there were, however, other non-Chinese members in the managerial cadre including a Filipino medical doctor, a Nigerian accountant and personnel manager. In terms of organisational division of labour and bookkeeping practices, the Chinese firm was the best organised. Managerial strategy in the Chinese firm was hierarchical and highly autocratic. The autocratic nature of the administration in this firm made the Chinese and the Nigerian personnel managers who executed disciplinary decisions very unpopular with the factory workers. Employee dismissals and strikes were most rampant in this firm. The dismissals[139] were characterised by tensions because the victims were often not eligible for entitlement or work benefits. Most employees saw their Chinese employers as 'less empathic' to subordinates and 'highly tenacious'.

The power distance between Chinese managers and their non-family employees is a general characteristic of most Chinese family firms. According to Silin (1976) and Redding (1989), the authority patterns in Chinese firms are borne out of Confucian work ethics, which tend to emphasise superior-subordinate relations. Confucianism 'sets forth its moral imperatives as duties rather than rights' (Fukuyama, 1996, p. 284). The language barriers further compounded the power distance in the Chinese factory. Chinese business operators in Kano and in parts of Nigeria are often handicapped by their inability to express themselves in both the local language and in fluent English. Some common tensions in the Chinese factory were therefore the result of communication problems and cross-cultural misunderstandings.

In contrast to the Hausa and Lebanese plastic firms, the Chinese management permitted the formation of a labour union in its factory. The presence of an employee union in the factory apart, the workers in the Chinese firm had less autonomy than their counterparts in the Hausa and Lebanese plastic firms. The Chinese firm was the only case among the three plastic firms in which employees from other parts of the country outnumbered the indigenous workers. Intra-employee relations on the shop floor were characterised by ethnic tensions and suspicions between local Muslim employees and their Christian counterparts from the Southern and Middle-Belt areas of the country. Perhaps due to the four extra working hours, employee remunerations in the Chinese factory were higher than in the Hausa and Lebanese factories. Nevertheless, at N1,300.00, the average monthly salary was hardly at par with the officially prescribed minimum salary in the plastic industry. The non-payment of past employee arrears by the Chinese management from a previous nationwide salary increment in 1992, was the biggest source of discontent for employees of the Chinese plastic factory.

Clientelistic Relations as Recruitment Channels

The survival and growth of a business enterprise depends on the productive partnership of both employees and management. Skilled personnel often exert leverage on the organisation. No matter the level of technology applied in a firm, incompetent or unwilling labour may contribute to failure of the business enterprise. Research on organisational performance has shown that people are 'the main engine' of any enterprise not just 'interchangeable parts of the corporate machine' (Waterman, 1987, p. 104).[140] Entrepreneurs and mangers who ignore the importance of the human factor do so at their peril. The relevance of recruitment is that it makes the provision of several candidates, which may be large enough to enable an organisation to select the needed qualified employees (Stoner, and Freeman, 1989). Clientelistic relations, which are predominant in Kano's business environment, influence entrepreneurial recruitment patterns.

Recruitment is used in this text to show the process of selecting or sorting out qualified candidates or applicants for specific job positions and eliminating unqualified applicants from the labour pool. In Nigeria, the sources from which labour can be recruited include universities, technical schools, and the unemployed labour market. Ideally, recruitment from the labour market is based on ability or suitability rather than affinity or social background. The framework for recruitment involves two important levels. A recruitment need may arise through labour analysis of a firm and the identification of areas requiring reinforcement of human input. Furthermore, the introduction of new technology or diversification in an enterprise may require workers with new skills to handle key positions.

In smaller firms, information on job openings is kept secret, and made available to siblings, neighbours, townsmen, and friends before formal advertisements are made public. Thus, newspaper advertisements in Kano and parts of Nigeria only help to give the appearance of organisational fairness. The reason or motive behind this is to allow relatives and ethnic members access to strategic positions in self-operated businesses. Most prospective applicants in Nigeria therefore see newspaper advertisements and formal recruitment exercises as mere formalities. Employee recruitment through newspaper advertisements was generally low in the Hausa and Lebanese firms, with both firms recording approximately three and two percent, respectively. In the Chinese firm, workers in the supervisory category, representing up to 15 percent, were recruited through newspaper advertisements. My investigations revealed that consulting firms were not patronised in any of the plastic factories.[141]

In recent times, the use of consulting firms to recruit employees is becoming increasingly popular among employers of labour in the larger multinational firms. Consulting firms in Nigeria are specialised in the recruitment of professional workers. The reasons for low managerial patronage of newspaper and consulting firms are probably threefold. First, consulting firms are quite unpopular with smaller firms because they tend to charge prohibitive fees of up to 25 and 30 percent of the first-year salary of the employee. Second, such firms tend to emphasise merit, which the operators of small indigenous businesses do not necessarily cherish. Third, because of the simple technology used in the plastic industry, the requirement of highly qualified persons, which might be obtained through the aid of consultants and newspaper advertisements, is often unnecessary.

Recruitment through clientelistic or social networks continues to be the major channel of hiring employees in the factories of Kano. Social and clientelistic networks include family, primordial and ethnic members, business associates, friends, religious and other casual acquaintances with persons in privileged positions. The figures show that 59 and 41 percent of the employees in the indigenous Hausa and Lebanese firms were recruited through social channels. In the Chinese firm, the employees recruited from social networks made up 55 percent of the factory work force.[142] Significantly, most employees in the Lebanese factory, forming 57 percent, were hired directly at the factory gates. Most of these were casual or target workers who drift daily from one factory to another in search of jobs. In the Chinese and Hausa plastic factories, employment from this source made up 30 and 38 percent of the work force, respectively. Direct hiring of labour was lowest in the Chinese firm. Some disadvantages with these methods of recruitment are that sometimes very unqualified and unreliable persons are recruited. Furthermore, clientelistic and social networks tend to generate feelings of security, complacency, negative work attitudes, and unproductive behaviour among employees.

There was an absence of training programmes for the plastic factory workers. Most of the newly-recruited employees in the Hausa plastic factories were not conversant with their roles or job descriptions.[143] This often led to reckless mistakes and raw material wastage on the shop floor. Employee training is an area that has persistently generated controversy among indigenous business entrepreneurs and the workers they employ. Small business operators often think that staff training is a luxury that only multinational and well established companies can afford.

Several reasons account for entrepreneurial considerations of employee training as an incursion into business profits. First, training usually entails huge financial budgets and expenses. Second, training also implies that, in the future, skilled personnel will tend to ask for promotions and higher wages. Third, in the labour market, highly competent and well-trained personnel in small firms are frequently lured by bigger firms with more attractive offers; this is a scenario small business operators try to avoid. Without training, recruitment emphases are placed on applicants' previous experience and qualifications to minimise training costs. Despite the recruitment of workers through social networks and at factory gates, a significant proportion of the employees in the plastic firms, representing 43 percent, had previous working experience in comparable firms.

Operators of small firms in Kano seldom patronise free training programmes offered for factory employees by the Industrial Training Fund (ITF). In nearly all the firms, management saw training channels as 'holiday resorts' and 'breeding grounds for radical ideas' among workers. Managerial neglect of employee training was a major source of discontent among the rank and file in the three plastic factories. In addition, the payment of poor wages is another area that has generated a considerable amount of controversy in recent years. Of particular interest to the analysis in this text, are the peculiar effects of clientelistic and social relations of production on union membership and bargaining matters in these factories.

Wage Bargaining in Non-Unionised Firms

The subject of wage bargaining is central to understanding entrepreneurial behaviour towards workers in a developing country like Nigeria. Collective bargaining is a process through which collective decisions or agreements are reached between employer and employee representatives. These agreements involve the application of a continuous set of rules that govern substantive and procedural matters with guidelines on how working relationships should be conducted (Windmuller, 1987, pp. 3–16). Collective bargaining is thus an industrial form of negotiation.[144] It sets out the procedures by which wages may be fixed or industrial grievances may be solved. The ideal purpose of collective bargaining is to provide a forum for decision making between workers and their employers. Essentially, therefore, collective bargaining becomes a rule-making process, a system of countervailing power in worker-management relations (Flanders, 1969, p. 149).[145]

Although the term, 'collective bargaining' applies to larger groups of employees, individual bargaining is recognised by the Nigerian Labour Law at the point of recruitment. By law of the contract, the individual employee is free to bargain with management on terms and conditions of employment. The problem with individual bargaining, however, is that a single request seldom carries enough weight to compel employers to change important decisions. In the non-unionised firms, the prevalence of individual bargaining and managerial wage discrimination among workers performing comparable tasks are among the causes of employee frustrations and industrial unrest. According to Nigerian labour provisions, all employees of the plastic industry are covered by bargaining agreements reached between employers and the National Union of Footwear, Leather, and Rubber Products. Paradoxically, bargaining agreements do not apply to the majority of the employees in the relatively small non-unionised plastic firms. Consequently, a most recurrent complaints received from employees in the three plastic firms was the issue of managerial neglect of bargaining agreements.

The nature and form that collective bargaining assumes depends on a country's level of political, socioeconomic, and cultural context in which it is conducted. Lester (1957, p. 401) contends that the noteworthy fact about collective bargaining is that it differs in many ways from one country to another and within a particular country from one period to another. This implies that, even within a specific country, the nature of collective bargaining depends on the state of the economy. In difficult situations, the rules governing collective bargaining may be bypassed or not applied at all. Collective bargaining often varies across economic sectors. In the private sector, industrial bargaining may differ from plant bargaining. While industrial bargaining usually applies to all companies and firms in the same industry, plant bargaining often has a narrow and restricted coverage. The outcomes of plant bargaining seldom extend beyond the plant.

Nigeria operates the *laissez faire*, or voluntary, system of collective bargaining inherited from the British colonial government. Fashoyin (1980) observes that, while this doctrine may have worked in Britain, it has clearly failed in Nigeria because of frequent violations by employers of labour. The failure of government in the implementation of bargaining agreements has served as an impetus for labour abuses in the private sector. Commenting on this unusual trend, Kilby observes that:

> In failing to fix the wages of its own employees by negotiation with employee representatives, the government had de facto precluded the operations of the voluntary system for half of organised labour. And indirectly, the quantitative and normative power of its example, the actions of government have done much to undermine the potential for collective bargaining in the private sector (Kilby, 1969, p. 298).

In spite of a lack of accurate figures on employee coverage on bargaining matters, it has been observed that this is very low and disappointing in most developing countries. The picture however, appears different in the industrialised parts of the world. In countries such as Sweden, Belgium, and Austria, collective bargaining covered almost all employees by 1980. During the same year, Germany achieved 90 percent, Netherlands 80 percent, Great Britain between 65 and 70 percent, and in the United States between 25 and 30 percent.[146]

Wage determination, on the other hand, allows employers to fix employee wages and remuneration. This may be by piece rates or monthly. In another context, wages may be fixed according to employee productivity.[147] However, employee productivity is difficult to assess, especially in a manufacturing process that involves the joint participation of many people. A common theme in industrial relations research is the proposition that 'a stable internal pay structure' is a precondition 'for a compliant and cooperative workforce' (Brown, and Nolan, 1988, p. 351). The current harsh economic climate in Nigeria and the accompanying high cost of living continue to fuel agitation for wage increases by employees who are entirely dependent on wages for their livelihood.

Although no direct links can be established between wages, salaries, and productivity, it is expected that a good wage, with its attendant positive motivation can be a reason for which a worker's performance is enhanced.[148] Out of the 166 employees interviewed in the three plastic factories, 133, or 80 percent, agreed that good salaries were their main motivators for hard work. The role of a prevailing idea 'of social justice and fair wages' can be understood only by considering the procedures through which wages are fixed. Poor, reckless and arbitrary fixing of salaries and wages across comparable firms have become means of generating industrial unrest in Kano and parts of Nigeria.

With the new economic developments, the most important aspect of industrial work, or *aikin Bature*, is the end-of-month salary, or *albashi*. Wages have become the prime determinants for accepting employment under the new economic system. Wage issues 'have inevitably loomed large in Nigerian industrial relations' not just because they are indexes of social status but also a 'determinant of the combination of goods and services' and comfort 'that can be afforded' (Otobo, 1987). In Kano and across Nigeria, employee unrest, and often-violent approach to, wage matters are the result of the tremendous social pressures they have to cope with at home.

There was no uniform structure for the determination of a national minimum wage until after independence. The earliest form of joint negotiation occurred with the establishment of wage committees in 1939 by the British colonial government (Yesufu, 1962). Wage committees were composed exclusively of government officials. Wage determination was therefore a unilateral issue in actual practice. Workers or their representatives were never a part of the process of the deliberations on wage issues. This situation improved in 1942, and was expanded to include workers' representatives after a series of agitations (Fashoyin, 1980, p. 105). The inadequacy of employee representation in bargaining was revealed by a nationwide general strike in 1945. In a bid to introduce equitable representation and a sense of fairness, the government introduced the British pro-type Whitley Councils in 1948. Though national in scope, they failed to make any meaningful impact after only one year of their application. Several wage advisory boards which were established by the government in the 1960s and the subsequent Adebo (1970), and Udoji (1974), wage commissions, and the Productivity Prices and Incomes Board (PPIB 1976), have been fraught with failures.

Thus, the history of collective bargaining and wage determination in Nigeria to date has been an unpalatable one. Wage disparities across comparable firms in the private sector have always been the sources of industrial conflicts. The role of government in the Nigerian bargaining process can be understood within a framework of a patrimonial role which it plays as a general overseer and controller of economic transactions. In most confrontations with employers in the private sector on wage issues, it often becomes apparent that workers are locked in confrontation with the state.[149]

Wage fixing in the three plastic factories was a unilateral decision often taken by managers. Conversely, workers were never consulted on wage issues. This procedure was a great source of discontent, especially for the unionised workers in the Chinese factory. The absence of employee unions in the Hausa and Lebanese firms resulted in divisions among employees in these factories, which denied them the opportunity of presenting a strong front on collective bargaining matters.[150] Andrae and Beckman (1998, p. 148) observe that, 'the personalised mode of recruitment in Kano reflected a management and enterprise culture where personal relations, and hope for protection matter'. Nevertheless, such relations, they observed constraint employee ability of joining collective actions against management.

Most of the collective agreements often reached between employers of labour and their employees in the non-unionised firms were neither implemented nor considered binding by factory management. The ultimate effects of the breaches of collective agreements were the manifestations of grudges and discontent over conditions of work and services, which became an integral part of the industrial labour process in these factories.

Entrepreneurs and Factory Workers

The expectations of many Nigerian workers in the modern factory often result in shock. The individualistic nature of plant and shop-floor relations and the relative lack of freedom contrast sharply with the free intimate relations of the larger society. In particular, the impersonal nature of human relations in the factory and the enforcement of rigid rules by supervisory and managerial staff to ensure that operations thrive and flourish are the initial challenges the worker has to adjust to. New employees in manufacturing firms are therefore required to conform to the organisational idea of 'fitness'. However, due to cultural differences, employee definitions do not often conform to modern organisational requirements. Personality differences also make certain individuals fit into participatory groups more easily than others (Vroom, 1960). Employee adjustments in the factory are essentially based on the pace prescribed by factory managers and not on individual coping abilities.

Most business managers do not condone lazy attitudes because they are a hindrance to productivity and an encroachment on profits. Individual differences and modes of adaptation to new environments often result in personality clashes with organisational goals.[151] Job retention therefore depends on the employee's ability to adapt to formal organisational rules. In many Nigerian factories, new employees often undergo a probationary period of six months. Managerial impatience with slow-paced workers is based on fear of wastage and depreciation of machines that are considered as additional costs to production. In the Chinese plastic company, the inability of most indigenous workers to cope with Chinese strenuous work demands led to many job losses.

The 'relational' and the 'practical' aspects of the organisational life described above refer to the labour process. Variations in the level of technology and managerial practices from one firm to another can lead to differences in the labour process. Furthermore, changes in the general economic climate and dynamics in the labour market often reflect on the labour process. For Burawoy (1982, p. 15), production relations are essentially shop-floor relations. Dimensions of shop-floor relations involve the relationships among workers, supervisors, and managers. The practical aspect of the labour process relates to workers' interaction with machines and the transformation of raw materials into semi-processed or manufactured goods.[152] In all these relations, there is the potential overt or covert discontent that arises from tensions over the control of machines, unprofessional attitudes of work mates, and subordination to others in the production process (ibid.).[153]

The labour process, which embraces a whole gamut of issues such as job security, management-worker relations, and opportunities for career advancement, can change under different management, technology, and training (Lubeck, 1986). Thus, while some technical equipments or machines may allow room for manoeuvre on the shop floor, others may require the full attention of their operators to prevent wastage or other undesirable consequences. In the plastic firms, the noise of the machines made conversation between workers on the shop floor totally impossible during working hours. Factory workers also did individual tasks that separated them from each other. Apart from the short break hour, the factory workers had no access to any form of social interaction. The repetitive nature of the tasks performed by workers made them vulnerable to boredom and alienation.

In the circumstances where managers and supervisors lack adequate technical skills, industrial labour may be over-stretched or overburdened with tasks that may not be well-executed or accomplished with prolonged delays. Sometimes the entrepreneurial quest to provide jobs for unemployed relatives, regardless of the firm's needs, may lead to overcrowding in machine rooms and on the shop floor. This problem was particularly common in the Lebanese plastic factory, where the ownership desire to provide employment to relatives and friends led to an overcrowded shop floor. Not surprisingly, the majority of the shop-floor employees in the Lebanese factory could not define their roles. The absence of clearly defined roles in the factory, generated tension within the rank-and-file of workers on the shop floor because of the feeling among the more responsible workers that some of their colleagues were idle hands.

Most of the factory workers in Kano, and in Nigeria more generally, find factory procedures difficult to follow, not necessarily because they are nonconformists, but mainly because the procedures often contradict cultural norms. The hierarchical arrangement of authority in the modern organisation requires that employee grievances are channelled through the immediate superior. This therefore implies that the ordinary worker seldom has direct access to management, in contrast to the larger society where individual grievances can be directly channelled to the superior. In the indigenous Hausa firm, where the managerial style was patrilineal, the organisational hierarchy was frequently bypassed. Employee access to management thus contributed to their contentment in spite of the incredibly low wages.

Religion and Shop-floor Relations

Nigerians, regardless of their social or ethnic backgrounds are generally very religious people. Religious values therefore permeate all social and economic boundaries. In the factories, religious and cultural values often become an integral part of the organisational behaviour. Due to the volatile nature of religious issues, resulting from a new wave of religious revivalism among the various religious factions across the country, both local and foreign business proprietors try to avoid meddling in religious matters. However, in their bid to boost employee morale and still be able to exercise effective control over them, business proprietors often try to fulfil certain basic religious conditions for their employees. Such conditions include managerial permissiveness towards religious worship in the factories and the provision of loans to employees during religious festivities.[154]

The population of Muslim employees in the three factories was highest, constituting 56 percent. Christian employees, who made up 42 percent of the employee population, closely follow this figure. A breakdown of the Muslim population further reveals that, those with *Sufi* links made up 43 percent, while *Sunnis* represented 13 percent of the work force. Employees of Catholic faith, who formed 19 percent, were in the majority among Christian employees. Protestant and local Christian denominations made up 12 and 11 percent of the employee populations respectively. Followers of traditional religion were a minority, representing only 1 percent of the entire labour force in the three plastic factories. A rather small proportion of the employees, making up 1 percent, gave no response about their religious affiliations.

There were, however, variations in employee religious affiliations in each firm. Whereas Muslim employees were in the majority in the Hausa and Lebanese factories, in the Chinese firm, they were the minority group. Interestingly, most Muslim Hausa, Fulani, Kanuri, Marghi, and Babur employees were of northern backgrounds, while their Christian Ibo, Calabar, Ibibio, Edo, Ijo, Gwari, Tiv, and Birom colleagues were mostly from the southern, delta, and middle-belt parts of the country. The Yoruba employees from Southwestern Nigeria were fairly equally represented in both the Islamic and Christian religions. Significantly, Muslim and Christian factory employees who together formed approximately 98 percent of the labour force, did not see factory work requirements and procedures as contradicting their religious and cultural values.

The first chapter mentioned that the proliferation of mosques in the city wards and factory yards of Kano makes the impact of Islam ubiquitous. Although there were other non-Muslim religious adherents in the factories, the religious activities of Muslim employees were more dominant because of the routinised and congregational way in which they were conducted. Sometimes the tensions in the factory-yard or shop-floor were products of religious misunderstanding of verbal utterances and behaviour from fellow workers considered by others as blasphemous or sacrilegious. Freedom of religious worship existed in all the firms, including the Chinese factory, where the Chinese management refused to comment on religious issues. In the Hausa and Lebanese plastic firms, where the proprietors were Muslims, the daily interactions in the mosque between them and their employees helped to create smooth working environments.

Many local factory employees, or *leburori*, as Lubeck (1986, p. 151) observes, are local Islamic scholars, or *malams*, who play significant roles in the community, such as running Islamic schools after factory hours. Among Muslim factory workers, the local *malams* who distance themselves from management are regarded as less manipulable and are therefore well-respected, despite the inferior roles in which they often find themselves due to their lack of Western education. A combination of Quranic knowledge and strong attachment to faith gives the factory *malam* a leadership role among other Muslim workers. In times of crisis, the factory *malams* and radical 'Islamic Brothers' are those looked upon for religious and other aspects of social leadership. During periods of tension and industrial unrest, employers of labour sometimes rely on the support and intercession of the influential *malams* within the ranks of their workers, to persuade their irate colleagues to return to work. Religious factors play a significant role in conditioning the social relationships in the business firms.

The social pressures on poorly paid factory workers, especially during major religious festivities such as the *Sallah* and Christmas, often mean that loan advances are sought from employers. Factory employees often rely on such loan advances during religious occasions to meet their many social obligations. This variety of needs includes clothing for spouses and children and ritual rams. In recent years, late or poor handling of *Sallah* and Christmas loans to workers have become sources of tension and unrest in most of the factories operating in Kano. Occasional gifts and loan-advances are strategies employed by both foreign and indigenous employers to ensure a sense of fairness, loyalty, and high productivity among their employees. Entrepreneurial and managerial treatment of employee requests during religious festivities constitute measuring rods for future reactions.

In some of the textile manufacturing firms operating in Kano, workers received six yards of cloth annually as take-home gifts to their families during the Islamic holidays, or *Sallah*. These generous gifts from textile industries triggered agitation for similar gestures in non-textile firms. Because of this, many employers of labour in non-textile factories have devised various remuneration strategies for their employees during religious festivities. There were, however, variations in the managerial remuneration packages in the three plastic firms. This occasional entrepreneurial and managerial largesse was extended to all employees, regardless of their religious backgrounds.

In the Hausa and Lebanese plastic firms, workers received *Sallah* and Christmas bonuses. However, while in the Hausa firm this was as high as half the employee's monthly salary, in the Lebanese factory, it amounted to less than a quarter. The Chinese plastic factory, on the other hand, developed a unique holiday bonus, annually slaughtering a cow for its employees during the Muslim festival of *Eid el adha* or *babbar Sallah*. However, contrary to expectations, the majority of the employees in the Chinese factory did not appreciate the gesture. In the Hausa and Lebanese factories sometimes long-serving and hardworking Muslim employees and relatives are offered sponsored trips to perform the annual pilgrimage, or *hajj* to Mecca as a way of boosting their morale. This practice sometimes makes some employees develop specific types of allegiance to their employers that make them vulnerable to manipulation. Concerning this practice, an indigenous foreman in the Lebanese plastic firm made the following observations:

> I have worked with the Lebanese people since I was a teenager . . . Now I am an old man . . . and although I have exceeded my retiring age, they still allow me to work to feed my family . . . A few years ago, the director of this company sponsored my trip to Mecca at his own expense . . . I had extra money on my return from this journey which enabled me to buy a plot of land on which my house is presently situated . . . I agree with the junior workers that the working conditions here are bad, but can I say that I have not been fairly treated?

Among factory employees, religious issues exercised an important role in their loyalty and trust towards the union leadership. Employee loyalty and suspicion of union leadership in the three plastic factories were linked to religious and ethnic background. It is worth mentioning, however, that intra-employee relations were neither as militant nor as polarised as depicted in Lubeck's (1986) study of Kano factory workers.

During the Islamic fasting month of *Ramadan*, working hours in all the factories were altered to suit the productive capabilities and religious activities of Muslim workers. The eight working hours in many firms were reduced to six. During *Ramadan*, most Muslim employees participate in the late afternoon public exegesis of the Qur-an, the *Tafsir Jalallain*, and other classical Islamic texts in the various wards of Kano. Because of the importance of this annual event to Muslims, factory owners often operate two hours less than their normal operation schedules. The evidence presented in this Chapter shows a direct link between cultural background and entrepreneurial ethics.

6 Entrepreneurs and the Profit Motive

> Profitability means, in the rational case one of two things: (1) the profit estimated as possible by ex-ante calculations, the attainment of which is made an objective of the entrepreneur's activity; or (2) that which the ex-post calculation shows actually to have been earned in a given period, and which is available for consumption uses of the entrepreneur without prejudice to his chances of future profitability. In both cases it is usually expressed in ratios-today, percentages-in relation to the capital of the initial balance (Weber, 1978, p. 91).

This Chapter provides an outline on the subject of profit. Much of the analysis in this Chapter is linked with the theoretical discourse on culture examined in Chapter One. Culture paves the way for entrepreneurial ethics and strategy regarding profit acquisition and eventual economic success. Entrepreneurial profit motives serve as stimulants for propelling the economic activities of Hausa, Lebanese, and Chinese business groups discussed in Chapter Five. Profit is the prime motive and reason for indulging in business activity. In the business sphere, the ability to earn profit is confirmation of entrepreneurial innovation.

Entrepreneurial and managerial perceptions about the ultimate outcome of a particular economic decision usually influence their response towards that activity. Not surprisingly, over 80 percent of the business operators interviewed in Kano cited profit as their main motive for being in business. Generally, profit is the motivating factor for entering business and often the basic reason for pursuing any line of economic activity. Entrepreneurial enthusiasm towards business often derives from the quest for profit acquisitions. Presumably, reluctance regarding involvement in any line of business activity stems from the fear of loss and failure. The inability of entrepreneurs to decipher a profitable outlet in a specific business activity is often the reason for their unwillingness to experiment.

Even in the face of losses, the inextinguishable faith in profit is a key factor that sustains entrepreneurs. Profit oriented opportunities are designed towards seeking 'powers of control over goods on a single occasion, repeatedly or continuously' (Weber, 1978, p. 90). There is a strong correlation between profit perceptions and risk-acceptance, discussed in Chapter Two of this study. Entrepreneurial and managerial perceptions on the ultimate outcome of a particular economic decision usually influence their response towards that activity. The functions of profit include rewarding the decision-maker, creating comfort and reducing anxiety. Profit can further serve as a monetary reserve for future capital investment (Keirstead, 1961).

Despite the crucial role of profit in influencing business decisions, there is no generally accepted theory on profit and profitability due to confusion over the terminology. The different analytical strands on profit show very little signs of convergence as the 'theoretical controversies remain intense' (Howard, 1983, p. 3). These analytical diversities and controversies stem from global economic transformations from smaller and simpler firms during the early years of industrialisation to the bigger and more complex corporations in the modern era (Keirstead, 1961). In the former, the entrepreneur controlled the entire business process, while, in the latter, there is an immense division of labour, and corresponding split of control functions. Whereas in commerce, profit may be easy to define or calculate, in manufacturing activity, assessment is often a cumbersome process.

In the classical text of Ibn Khaldun (1967, [1377]) mercantile or commercial profit is categorically subsumed in the notion of 'buying cheaply and selling dearly'. This conception of profit is somehow simplistic and medieval in outlook. Godelier (1972, p. 66) argues that the definition of profit 'as the excess of price over cost of production' does not go beyond the origin or effects of profit. In the modern manufacturing firm, profit is theoretically held as 'the amount by which the total revenues of a firm exceeds its total costs' (Keirstead, 1961, p. 547). For Adam Smith (1981, [1776, 1789]) profit is the amount that could be consumed without diminishing or dissipating capital. Thus, capital is a key determinant of profit in Smith's analysis of profit.

In his 'treatise of money' in the 1930s, Keynes observes that profit, which is the result of lack of equilibrium in an economic system, is 'an accidental windfall occasioned by temporary changes in price levels as a whole'.[155] Writing along Keynesian lines of imperfect conditions created by market forces, Robinson (1961) observes that profit is a product of the monopsonistic position in the factor market. A variant of the Keynesian disequilibrium theory of profit is the equilibrium theory, which has become known as the continental school.[156] Schumpeter considers profit as an effect of the disturbances continually created by entrepreneurs in the economic system.

Contemporary theories on profit as a product of system equilibrium assume 'perfect foreknowledge' of market conditions. In the works of Ricardo (1986, [1955]) 'profit is reward to capital or stock'.[157] Profit is generally held as surplus value in relation to all capital invested in the production of commodities (Godelier, 1972, p. 67). Despite the influences of classical over contemporary theories, there are hardly any agreements on common terms such as 'surplus' and 'profit'. The Ricardian theory about profit has capitalist development as its focus. It argues in favour of an inverse relationship between profit, and the low wages entrepreneurs pay their employees to enhance the retention of profit margins. Ricardo observes that 'the rate of profit was determined exclusively by conditions of production in wage-good industries' (ibid.).

Economists disagree on whether wage and salary payments should be included or excluded in profit analysis. Whereas wage payments are included in Sraffa's treatment of surplus, in the works of Ricardo they are not. In Sraffa's (1973, [1960]) analysis, there is an interdependent link between wages and profits. This implies that the 'rate of profit is not determined until wages are specified'. Following from Sraffa, wages and salaries in this text are treated as a cost of production separate from profit. It is, however, necessary to state that an in-depth analysis of profit from an economic perspective is beyond the scope of this study. Generally, economic abstractions about profit tend to adopt *ceteris paribus* assumptions and the retention of constant variables, which are analytically and empirically slippery from sociological perspectives. Other methodological limitations often arise because profits are by nature unpredictable. Accurate predictions of profits are not only problematic for analysts and researchers, but for business practitioners as well.

Definitions about business profit are based on accounting figures pertaining to records on the balance sheet that deal with profit and loss. In accounting statements, profit is taken as 'a surplus of assets over liabilities' (Weber, 1978). However, balance-sheet statements about organisational turnover sometimes tend to give false indications about business performance and profit. Accounting figures are based on certain conventions, which are not usually adequate measures of profit. Consequently, since measures are based on verifiable events and documents, profit is recognised only for sales and assets (Andersson, 1987, p. 63). Other limitations of the accounting method as summarised by Edwards and Bell (1972, pp. 10–11) include understating or overstating profit when gains or losses are still incomplete; distorting capital expenses by measuring them against depreciation and rent; and failing to recognise general price changes.

Despite the limitations of accounting figures, the fact that their weak areas are clearly known allows researchers to make meaningful interpretations (Andersson, 1987). In the light of this, the current operating profit (COP) is the approach used in this text. The current operating profit is based on consistent use of current values, which include; current sales, current cost of materials, current costs of wages, current depreciation on current value of assets, and current interest on borrowed capital. Business profit is therefore taken as current operating profit, and realisable cost savings during a specific period (ibid.).

Profit Perception in Business

In the introduction section of this chapter, the various conceptions on profit among economists, and business entrepreneurs have been examined. Ostensibly, there were variations in entrepreneurial, managerial, and employee perceptions regarding profit. Approximately 32 percent of the respondents, who were mainly entrepreneurs, traders and business managers, perceived profit as the difference between the buying and selling price. Although this definition is rather simplistic, the perception and calculation of profit in monetary terms conforms to the capitalist notion of profit (Rodinson, 1974).[158] Most of the entrepreneurs however, constituting 50 percent, perceived or defined profit in terms of additional earnings on capital investments. Only 9 percent of them defined or perceived profits as assets exceeding total costs or in terms of surplus of production, respectively.[159]

The Hausa conception of profit is often unclear. Usually, wages, revenues, and other costs are included with profits. Among merchants and traders in the commercial sector, profits are calculated on a daily, weekly, or monthly short-term basis. The ease with which commercial profit is calculated, is perhaps the reason less-educated individuals have a predilection towards sole-proprietorship. In dealing with manufacturing profits, several variables have to be considered. These include the cost of labour, energy, raw materials, transport, excise duties, value-added tax, and depreciation costs of machines. Conversely, profit calculation in a single venture is a less cumbersome procedure than in a joint venture, which requires profit sharing with partners. In spite of this, the profit-making ventures of Lebanese, Chinese and Hausa business groups covered by this study can be described as 'economic', because they were conducted in a peaceful manner. For Weber (1978, pp. 90–91), profit-making, can be described as 'economic as far as they are obtained through peaceful methods.

Islam encourages and legitimises entrepreneurial initiatives towards profit acquisition because profit is considered part of 'Allah's bounty'. However, Islamic economic principles emphasise that profits must have religious sanctity. The sanctification of business profits implies that the poor consumers in the economic system must be protected from the insatiable greed of entrepreneurs. The tacit assumption is that an investor's quest for profit must not destroy relations with God and other people. In Hausa communities, entrepreneurs who place profit acquisitiveness above religious and moral interests are often the targets of societal ridicule. Both customers and business colleagues sometimes shun such entrepreneurs. Therefore, excessive increases may be labelled as dirty profits or *kazamin riba*. Clean, or *halal*, profits are those seen to be earned within tolerable cultural limits, without excessive exploitation of the underprivileged. Writing about African economic traits, Dia observes that:

> African economic psychology is generally characterised by powerful connections between the objects, human and supernatural . . . The quest for equilibrium with other human beings and with the supernatural is generally the guiding principle . . . the main concern seems to be maintaining social balance and equity among the group (Dia, 1992, p. 176).

Islamic views regarding profit acquisitions within tolerable limits contrast with the limitless capitalist notion of profit. For many indigenous Muslim entrepreneurs, Allah's blessing, or *al-barka* on business activity, matters more than the profit margin. It is therefore generally held that only profits that receive blessings will be eventually sustainable. Indigenous Hausa entrepreneurial responses to profit-related questions are usually tinged with humility. This is because Islamic tenets encourage humility in the face of marginal profits or losses. Financial losses and economic setbacks are therefore regarded as temporary trials from Allah. In view of this, forbearance and humility are the essential attributes worthy of attracting divine blessings or *al-barka*, greater rewards, and future profits.

Most of the Muslim entrepreneurs in Kano will not admit business losses for the mere fact it may sound as a reflection of ingratitude to God. A common Hausa maxim that 'a person who underrates little earnings is a thief' further encourages entrepreneurial contentment towards all financial earnings, no matter how minuscule. Entrepreneurial responses to profit related questions often begin with 'thanks to the almighty Allah', prior to any explanations. In the commercial sector, Grégoire (1992, p. 146) observes that *Alhazai* gains and profits are difficult to evaluate because of the confidential nature in which their economic transactions are conducted. In many indigenous enterprises, profits are concealed by inadequate accounting and unrecorded withdrawals of cash or kind (Harper, 1984, p. 15). Similarly, profit earnings in foreign-owned enterprises, are sometimes artificially distorted by transfer prices in order reduce tax payments.

Determinants of Profit

No single factor determines profit in a business enterprise. Rather, several exogenous and endogenous factors and other impinging variables usually combine to make profit earnings feasible. This myriad of factors can further be categorised as short-term and long-term. The exogenous factors that determine profits are contingent on both macro and microeconomic events in the system. During times of recession or high rate of inflation, the prohibitive cost of raw materials and imported spare parts often overwhelms potential gains. In such situations, businesses stagnate. The costs of inputs to the manufacturing process therefore exercise significant influence on profits. Since a greater amount of inputs in the local manufacturing activities are imported, direct information and access to foreign companies for the acquisition of inputs also go a long way towards reducing production costs. The competitive advantage of foreign firms operating in Nigeria is their direct access to manufacturing inputs over their indigenous colleagues. The fact that indigenous industrialists often have to source inputs through second-hand dealers implies that their cost of production is higher than comparable foreign firms in the same line of manufacturing activity. The high cost of inputs is a significant factor that limits the ability of indigenous entrepreneurs to earn meaningful profits.

Depreciation in the value of the naira since the turn of this decade, and the high cost of production have been responsible for low profit margins in most Nigerian firms. The high rates of inflation favours speculative business in the service sector rather than productive manufacturing activities. Bureaucratic delays and sloppy implementation of macroeconomic policies by Nigerian governments are often impediments to business profits. Furthermore, political instability and intermittent changes in policy guidelines influence entrepreneurial forecasting, and consequently, limit their ability to increase profit. Other exogenous factors effecting profit earnings include poor road networks, persistent power failures, and inadequate water supply for factories.

Consumer behaviour is another external factor tied to business profits. The level of economic performance often affects the rate of employment, level of incomes and consumer purchasing power. Weber (1978, p. 93) observes that there is a correlation between the general purchasing power in an economy and the ability of investors to make profits. In a bad economy, low consumer purchasing power limits entrepreneurial abilities to make meaningful profits. When asked to recall any period in their activities when business was profitable, the Hausa, Lebanese, and Chinese proprietors of the three plastic firms unanimously affirmed that 1970s and early 1980s where their best moments in business transaction. Conversely, the recession years of the early 1990s were unanimously identified as bad years in terms of business profitability. The low business profitability in the early 1990s was due to the decline in the general purchasing power in the Nigerian economy following the world economic recession.

To achieve profits, indigenous Hausa entrepreneurs work tremendously hard in diverse business spheres, sometimes exceeding normal daily working hours. In their quest to earn profits, both entrepreneurs and managers sometimes have to adapt to a couple of flexible strategies. Rigidity and lack of innovation cause business losses and failure. Entrepreneurs often capitalise on the ignorance of buyers and consumers about prices elsewhere to ensure profits. Sometimes goods have to be hoarded and artificial scarcities created for earning profits. In commerce, and especially economic transactions involving agricultural produce, the patience of the seller and his/her ability to hold goods until lean period largely affects the profit margin.

Hogendorn's (1987) study of commodity exporting in Kano shows that the worst form of loss was the sale of agricultural crops by peasants while they awaited harvest. There appears to be a positive correlation between a trader's income and profits. Generally, traders with higher incomes can hold goods until prices appreciate. A common strategy among the merchants of Kano is the storage of goods until their prices are on the upturn. The practice of hoarding goods by wealthy merchants in Kano means that sometimes poor storage of goods leads to heavy damage and untold economic losses to the local business people.

In the industrial sector, the level of the prevailing money wage and entrepreneurial and managerial abilities to keep wages and salaries low also affect the rate of profits because wage increases generally encroach on profits and economic gains. The lack of employee access to wage bargaining, as shown in Chapter Six, allowed entrepreneurs to pay low wages in the factories.

A relevant determinant of profit is employee behaviour, especially regarding productivity. However, as far as the three plastic firms were concerned, employee attitudes did not pose problems to entrepreneurial abilities to make profits. The evidence in the three plastic factories has revealed that the exogenous factors played a stronger role in determining the rate of profits earned by the business proprietors.

Profit Management among Entrepreneurs

Whether written by indigenous or foreign scholars, the literature on indigenous entrepreneurial management or use of business profit in most developing communities is not very encouraging. Profits are not necessarily ends in themselves, and would be of no value to any economy if they were hidden under a mattress, held in a foreign bank account[160] or frittered away on conspicuous consumption. Available evidence suggests that this gruesome picture may not be entirely correct. Nevertheless, the manner in which business profit is managed or used is crucial for the success or failure of a business enterprise.

A significant proportion of business profits earned by both indigenous and expatriate entrepreneurs in Kano is reinvested in other economic ventures. A relatively large number of entrepreneurs and managers, making up 55 percent of the sample, maintained that they used profits to pay debts and salaries of workers. Approximately 30 percent showed that profits were used for buying more raw materials, while 11 percent maintained that, after using part of profits to defray production costs, the rest were saved for the future.

Entrepreneurial use of profits for solving social and personal problems scored lowest on the responses received, representing only 4 percent. The responses received do suggest that social problems did not take a toll on entrepreneurial profits. However, responses of this nature from entrepreneurs cannot be taken as entirely valid, since most of the social remittances from business profits are not recorded. On a general note, indigenous entrepreneurs are more likely to use their profits to solve local problems than their expatriate counterparts. Among Levantine business groups, business profit usually goes toward expanding or divesting into newly created market niches.

Informal sources from factory employees suggested that successful Levantine businessmen often extended loan support to new coethnic members to establish their own businesses. This practice further serves to boost the spread of Lebanese businesses in Kano. Given the limited number of Chinese businesses in Kano, very little can be said about their use of profits. However, among the proprietors of the Chinese plastic factory, there was diversification of profit into other fields such as hotels and restaurants, iron and steel production, textile commercial outlets, and confectionery manufacture. The Chinese were also very thrifty in the payment of salaries and allowances to their employees. Despite subjecting their employees to twelve-hour days, the salaries of employees in the Chinese firm were lower than those in comparable plastic firms.

Profit management among indigenous entrepreneurs is particularly worthy of notice. Muslim entrepreneurs engaged in industrial activities often confront two types of taxes. The first form of tax is distributed among the poor in the form of *zakkat*, while the other is paid to the state. To the practising Muslim investor, the religious tax, or *zakkat*, is most important. In view of this, 57 percent of the Muslim entrepreneurs and managers whom I interviewed affirmed their payments of the Islamic annual tithe or *zakkat* to the poor. In contrast to most Islamic societies, the payment of *zakkat* by the affluent in northern Nigeria goes directly to the poor rather than the state. Consequently, tax default to the state is considered as a non-issue so long as religious obligations are met. Entrepreneurs' attitudes towards official tax payments are strengthened by their low-level trust in the Nigerian tax system, which is mired by corruption and financial improbity.

It is often easier to conceal business transactions or to evade official taxes in the commercial than in the manufacturing sector. Commercial activity usually involves mobile and invisible assets, while in manufacturing, assets tend to be fixed and are often visible. Apart from the payment of *zakkat*, indigenous entrepreneurs also use profits, within culturally and religiously acceptable limits, to assist their immediate families, and primordial groups. Among the most active Muslim entrepreneurs of Kano, profits are expended on philanthropic activities, such as the construction of Islamic schools and mosques, entertaining visiting religious scholars, and financing religious trips of the poor to the annual pilgrimage in Mecca. Such unsolicited generosities are undertaken by merchants in anticipation of the blessing, or *al-barka*, of *Allah*, *sheik,* and *malam*. Today many indigenous Hausa entrepreneurs are founding, education, health institutions, construction roads and bridges from personal incomes for the sake religion and societal progress.

Some of the negative aspects of the indigenous entrepreneurial use of profit include the payment of poor wages to employees and the pursuit of socially and religiously disproved acts of extravagance. These are often exemplified in the funding of expensive social ceremonies and the ownership of a fleet of cars often including Hondas, Mercedes-Benzes, and Rolls-Royces. In the past, it was fashionable for a business person in Hausa society to ratify success through the publicity of praise singers or *maroka*. Today, the traditional forms of generosity for the sake of prestige are on the decline. The reasons are perhaps twofold. First, the religious reforms that have swept across Hausaland since the past three decades. Over the years, Islamic reformists have fought against extravagance in the local communities. As earlier noted, new forms of rationality and optimal use of profits have emerged among the affluent Muslim business community. Second, the current economic hardships have foisted upon indigenous entrepreneurs some amount of self-discipline.

Societal perceptions have far-reaching influence on the behaviour of entrepreneurs in economic activities. Usually, successful individuals and groups in traditional communities tend to show concern about what the community or public feels and says about them. Group interests tend to shape perceptions about the use of wealth derived from economic success. Sometimes, the pressure from peers and primordial groups and the desire to affirm membership in a specific group may lead to unplanned new economic undertakings and generous donations during social occasions. Indiscriminate marriages among affluent merchants make divorce rates in Kano very high. Callaway's (1987, p. 42) study shows that 35 to 40 percent of the married women in Kano have been divorced at some point in their lives.

Although it is rare to find a healthy adult male in the African traditional setup without a dependant, the dependency burden on wealthy business owners is relatively higher than with factory or wage earning employees. Conversely, while most factory workers, representing approximately 49 percent were single, all the entrepreneurs and business managers covered by this study were married. Moreover, due to the widespread practice of polygamy, 45 percent of the business proprietors were married to two or more wives, whereas only 8 percent of the entire population of employee respondents had two wives. Accordingly, there were variations in the number of children supported by entrepreneurs and their employees. While married factory workers with five or more children were less than 4 percent of the sample, approximately 25 percent of the business entrepreneurs had five or more children. Although no factory worker supported eight children, 9 percent of the business proprietors supported between eight and nine children. The statistical data confirm a strong correlation between economic status and family size or the dependency burden.

The subcultural deviants among the merchants of Kano sometimes indulge in the expensive maintenance of multiple homes, one in which the family is kept, the other where mistresses are hidden in the obscure *Sabon-gari* section of the city. Extravagant maintenance of mistresses and the pimps, or *kawalai,* who straddle between them with errands, sometimes encroaches on business profits of this category of merchants.[161] Nevertheless, because of the sensitivity of profligate private lifestyles in Hausa society, it has not been possible to obtain statistical data on this specific problem. Although this kind of deviant and nonrational behaviour was possible among syncretist and professing Muslims, it was seldom found among puritan and reformist Muslim business groups in Kano and elsewhere in northern Nigeria.

The economic implications of the uses of profit are very complex. Since indigenous profits are almost diversified in commercial activities and to a less degree on modern manufacturing activities, there is a stagnation of the industrial sector while commerce continues to swell. The inadequate supply of industrial goods to fill demand means that such goods have to be imported from the Asiatic countries such as Japan, China, South Korea, Taiwan, and Singapore to supplement shortfalls in local production. Some harmful effects of indigenous entrepreneurial investment policies are self-evident. Entrepreneurial emphasis on small-scale family commercial ventures has led to an inadequate supply of jobs to absorb the increasing population. An abiding feature of this scenario has also been the emergence of pockets of splendour amid squalid conditions and increasing numbers of city vagrants, or *Yan daba*. The coexistence of affluence with extreme poverty has given rise to perennial economic riots under the guise of religious or interethnic disturbances. These scenarios reveal linkages between socioeconomic contradictions and societal instability.

Profit-Sharing and Trust

Since most business will stop operating in the face of losses, workers tend to believe that the fact that a firm continues to operate is enough indication that its business is profitable (Liles, 1974). Most workers tend to have a rather false optimism about the profitability of the firms in which they work, even where such businesses are bankrupt or merely surviving. The reason is that, in most societies, perpetuity or continuity in business operation serves as a measure of profits. Entrepreneurial motivations and enduring partnerships with employees are contingent upon profit earnings. Since profit is the measuring rod for continuing in business, losses may lead to employee redundancies and retrenchment. Unlike entrepreneurs and employers of labour, employees perceived profits in a slightly different context. A shop-floor employee in the indigenous plastic firm affirmed this notion with a more rhetorical question:

> Of course!...I know this firm is making huge profits, or why will any business enterprise continue to operate for over ten years if it is not making profits?...I am very sure that the proprietor of this firm is not a fool when it comes to making money.

This rather optimistic view regarding profit was not a solitary one. Most employees in the three plastic factories were unanimous on this view. Approximately 98 percent of the 166 employees interviewed agreed that the firms in which they worked were profitable, while only 2 percent believed they were not. In answering a question relating to the signs or symptoms of business profits, 26 percent of the employees thought continuity in business operation was an indication of profitability. Approximately 49 percent of the employee respondents in the three plastic factories saw the size of a firm as an indicator of profitability. Consequently, 'bigness' was the basic and most relevant sign of profitability according to the plurality of the employee respondents. A machine operator in the Chinese factory provided a rather vivid description about his understanding of business profit in the following manner:

> There are no better measuring scales or instruments like the human eyes. The mere size of this firm should convince any skeptic that huge profits are being made here, or what else can an ordinary worker like me say? . . . The volume of activities in and outside this factory is enough proof to substantiate my claim about profits.

It is pertinent to mention, however, that due to lack of access or knowledge of factual figures about organisational performance, employee views and hunches about business profits could be spurious or very misleading. Employees and outsiders often draw fallacious conclusions that all big firms are profitable corporations. Experience has shown that the size of a firm is not always correlative with profits. In reality, sometimes smaller and medium-size firms earn profits at the expense of bigger companies. The differences, however, may be in the profit margins. Very often, the profit margin in bigger firms is larger than in smaller and medium-size firms.

Employee optimism and convictions about profit contrast sharply with those held by their employers. Of the total number of 44 entrepreneurs and managers interviewed, only 13 of them, representing approximately 30 percent agreed that the businesses they operated were profitable. A larger proportion of the entrepreneurs, representing 59 percent, saw the businesses they operated as unprofitable ventures. Generally, employee and entrepreneurial conceptions about profit cannot be taken as given since they represent diverse interests. While workers may use certain parameters or indicators to push for wage increases, entrepreneurial reluctance to admit profits may a response to fear of employee unrest and agitation for improved working conditions. Business proprietors are generally reluctant to admit profit because of the feeling that such openness may expose them as profiteers and render them vulnerable to higher taxes from government officials. Sometimes, joint ventures may even be avoided due to entrepreneurial distrust about the modality of profit sharing.

For most employees, entrepreneurial profits often serve as a boost for further agitation for higher wages and better working conditions. Profit-centred issues and employee motivation are therefore highly complex issues. For factory employees, wages are never high enough to meet a considerable amount of obligations at home, whereas for business entrepreneurs, employee demands for wage and salary increases are encroachments on business profits. The widening chasm and bifurcation of interests on profit sharing often lead to distrust and friction in the working relations. Profit sharing is therefore the most relevant cause of disagreement between business owners and their employees in modern firms. Depending on the way business profits are handled, they may become a source of motivation or discontent among industrial workers.

In the three plastic firms, managerial payment of poor salaries and wages was a source of discontent and distrust among employees. Distrust was particularly high among low ranking and poorly paid employees who believed that the firms in which they worked were profitable. Approximately 80 percent of the employees in the three plastic factories saw good salaries and wages as prime motivators. A small number of employees representing 20 percent, identified 'self-esteem and recognition by the employer' as motivating factors.

The evidence in Kano and other parts of Nigeria has shown that 'good salaries and wages' are major preoccupations of factory workers. As will be expected, 80 percent of the workers in the three plastic factories wanted business proprietors to pay better salaries. When viewed from the angle of a hierarchy of needs as articulated by Maslow (1954), physiological needs continue to dominate the interests of most Nigerian workers. Consequently, a significant amount of labour unrest in Nigerian factories derives from employee agitation for better remunerations. When asked if they aspired to work at their current jobs until they retired, not one employee among the respondents in the three plastic factories answered positively. It thus appears quite strongly that most of the workers in all three firms were not satisfied with their working conditions.

Entrepreneurs and Employee Trust Relations

Economic relations of production from a Marxist perspective are generally asymmetric. The relationship between entrepreneurs and employees in the work environment can be asymmetric because of the differences in the exercise of power and control of financial resources. A low level of trust frequently characterises asymmetric relationships. Mutual trust can, however, be present in relations of power asymmetry where there are equitable rewards and punishments, etc. Managerial compliance or non-compliance of industrial agreements is particularly strong in reinforcing or destroying trust relationship with junior workers. Trust relations between many employers of labour and factory workers in Kano is low because of the failure of the former to fulfil legal obligations, agreements and terms of employment. In the Chinese Plastic factory at Bumpai, the evidence suggests that managerial breach of an industrial agreement with the factory employees was responsible for the violent upheavals and strikes in the factory. In many Nigerian firms where entrepreneurs do not compensate employees for injuries sustained on the job, and refuse to pay retirement benefits, industrial relations are tinged with mutual suspicion.

In manufacturing firms, trust is a key factor responsible for the recruitment patterns in indigenous and Lebanese firms. To trust an employee or applicant, sometimes entrepreneurs need factual knowledge about their history or background. The emphasis on employee recruitment through clientelistic and social networks stems from a notion of 'guaranteed trust'. This implies that, in employee malfeasance or breach of trust, there is always a reliable guardian to be contacted or held accountable. Entrepreneurial recruitment patterns have implication for trust. In the indigenous Hausa factory, the predominance of Hausa and Muslim workers created suspicion between Muslim Hausa and non-Muslim Hausa employees from the southern parts of the country. Non-Muslim workers were sometimes suspicious of their colleagues whom they suspected could pass certain information about them to management. Likewise, in the Chinese factory, the predominance of Christian employees in a predominantly Muslim society sparked distrust among local workers, who saw the measure as a deliberate ploy to weaken the strength of indigenous Muslim workers.

The absence of deep trust in employees in the plastic factories underlies the fact that very little investment is made in their training. Among expatriate business operators in Kano, the fear of loss of business control accruing from the partnership with powerful local classes, is a strong reason for the handpicking of weak partners often known as 'sleeping directors'. Foreign business operators also want business partners with reliable identities and local influence.

In the factories, monitoring is applied to control untrustworthy individuals (McAllister, 1995). Dahl (1998) for instance questions the rationale behind the employment of untrustworthy employees, because in his view it is counterproductive. Anyhow, distrust in employment relations could reduce productivity or encourage sabotage in the manufacturing process. In both the Chinese and Lebanese plastic factories, managerial low trust in employees resulted in the close monitoring on the shop floor. Close monitoring in these factories yielded tensions in the shop-floor relations between machine operators and line supervisors. The stringency of controls especially in Lebanese and Chinese firms, and the absence of employee autonomy negated trust and heightened tensions between factory owners and their employees in these two factories.

For North and Thomas (1973, p. 1), 'efficient economic organisation' has been the key to growth and development in many parts of the world. Polities specify and enforce property rights. Institutional constraints need to be restructured in order to redirect incentives (North, 1996, p. 112). Part of this rest in the ability of policy makers, especially the government in ensuring a functional industrial relations system. A corruptible government inevitably constitutes an institutional handicap. The absence of efficient organisations, combined with poor wages, job insecurity, and the disappearance of career paths in manufacturing firms, compels many Nigerian factory employees to seek self-employment in the informal sector. Distrust for factory employment implies that most industrial workers in Kano do not aspire to be entirely dependent on wage employment for very long. Employee aspirations for eventual self-employment make them favourably disposed to technical, mechanical, and professional roles such as welding, panel-beating, electrical work, carpentry, vulcanising, and painting which can be easily used outside the factory.

Informal loyalties and trust relations sometimes do occur among the rank and file of the junior workers in the shop floor. Employee trust relations are important because they are often avenues for covering the minor offenses of colleagues such abstentions and moonlighting on the job. The relevance of informal trust relations in the work environment is that they serve to provide mutual benefit to employees. Informal loyalties among junior employees in the three plastic factories were strongly influenced by ethnic background, place of origin and religion. This however, does not imply that there are no transethnic trust relations or across sectarian lines. On the contrary, these do exist to a significant degree.

Part of the low trust in employees is reflected by the refusal of employers to recognise unions or employee associations in the local Hausa and Lebanese plastic factories. All the employers of labour covered by this study viewed workers' unions as unreliable and problematic. Reward patterns and organisational trust relations are interlinked. Arbitrary rewards to employees working under similar conditions have negative implications for trust.

Among local entrepreneurs, the desire to ensure regular profits with a combination of cultural, political and economic factors produce specific business patterns and structures across northern Nigeria and many developing societies. Structurally, many businesses are characterised within the much narrower domestic and peripheral contexts in which they operate. The form of ownership, size and hierarchy of decision-making process, are largely determined within the family boundary.

Family Businesses in Northern Nigeria

The strong sense of the family in Hausa society is reflected by its relative cohesion. Observation of the businesses in Kano and parts of northern Nigeria shows that most of them are commercially-oriented family firms run primarily by adult males. Family ownership and control also dominate Lebanese firms, which compete side by side with most indigenous firms. The reasons for the dominance of family and patrilineal control of businesses are both cultural and religious.

Lovejoy's (1973) study of 150 cola traders in Kano reveals that 129 of them were grandsons of cola traders. The basic rationale behind the family business, is the provision of employment to kin members while limiting outside interference with and control of, family resources. Lovejoy argues that many business enterprises do not outlive their founders because, under Islamic law inheritance is partible (ibid.). In spite of their transient nature, Yusuf's (1975) study of Kano businesses demonstrates that those large family firms with sufficient capital base have survived into second generation enterprises. Some of these large indigenous family businesses have survived because very often children want to continue in a line of economic activity where their parents have been particularly successful.

Today, the main agents of economic transformation in Kano include Nigerians, Lebanese, Syrians, Asians, European, and American multinational firms. In recent years, any discussion about economic activities in Kano would be incomplete without a tribute to the contributions of the Ibo and Yoruba business communities. These groups from the Southeastern and Western part of Nigeria are no longer peripheral, but in the mainstream of Kano business. The Ibos of Kano, in particular, control extensive spare parts and building materials trade. The Yorubas, on the other hand, are most active in paper manufacturing and printing, tyre retreading, and the transport service sector.

Levantine investment in Kano particularly dates from the precolonial era. However, its impact became more noticeable after the First World War, when the Lebanese entrepreneurs obtained licence as both importers of European goods and exporters of agricultural commodities (Olukoshi, 1985, p. 54). A considerable amount of Levantine investment is in the area of textiles, pharmaceutical and other small scale manufacturing, and food-processing activities. Levantine and Indian control of the textile-manufacturing sector is substantial, while indigenous Hausa involvement in commerce is overwhelming. Some active Levantine business groups in contemporary Kano include P. S. Mandrides, Hassan Brothers, Raccah, Fawaz, Shour, Fahid and Yassin. There is now an increasing presence of Asian investors, of especially, Indian, Chinese and Japanese origin, in the textile and plastic manufacturing sectors of the economy. The impact of multinational firms in Kano is, however, not as significant as in Lagos and Kaduna. The industrial sector in Kano is therefore oriented towards light consumer goods, a feature that contrasts with heavy and technical industrial sectors of Lagos and Kaduna.[162]

On the local front, a relatively large number of indigenous businesses such as Dan-Tata, Gashash, Sani Marshal, Issiaka Rabiu, Baba Nabegu, Nababa Badamasi, Garba A. D., Dan-Kabo, and Dan-Gote continue to dominate the arena of economic transactions. The primary reason for the predominance of petty commercial activities in Kano is that they often require very little capital to start. Investments in commerce, which can often be recouped in less than one year, also provide quicker profits.[163] In spite of this, some leading indigenous Hausa industrialists including Dan-Tata, Gashash, Sani Marshal, Issiaka Rabiu, Nababa Badamasi, Garba A. D., and Dan-Gote have substantial capital investments in commerce and industry which often exceed Levantine and Asian investments.[164]

The largest single industrial investment in the mid-1980s of N12 million was by Nababa Badamasi (Forrest 1994, p. 20). Apart from his investment in the Gaskiya textiles, Badamasi also operated a sweet factory, Moon Confectionery as early as 1972. Other investments by this local business man include large shareholding in Raleigh Industries, Funtua Cottonseed Crushing Company, and large equity participation in the Nigerian Industrial Development Bank (ibid.). The boom in aviation transport in Nigeria during recent years has led to substantial investment in local air shuttle services by the business community.

Most Hausa businesses, as earlier observed, are family owned. Weber (1951, p. 237) recognises the family as an obstacle to rational capitalism. He contends that the family structure and especially the 'sib' throttled capitalism in both China, and India. Family firms are considered as inappropriate for the new conditions of industrial development (Lazonic, 1991). Coleman (1987) counterpoises the traditional image of the family firm by asserting that family and nonfamily firms exhibit similar behavioural characteristics, which reflect societal culture. According to Kets-de Vries (1992) the notable problems inherent in family enterprises include matters of succession, raising capital for further expansion, appointment of incompetent persons to senior positions, extraction of excessive resources from the firm to support uneconomic family lifestyles, and the extension of family conflict into business.[165] In order to ensure a regular flow of income and to improve the living standards of their families, owners of family enterprises often do not remain reliant upon a single economic activity. Small family businesses tend to be weak in generating employment (ibid.).

Despite these limitations, the impact of family values on economic life poses a complex and contradictory picture. The crux of the matter is that, while family businesses may fail under certain conditions relating to particular individuals, groups, cultures and the environment, they may be successful in others. It is therefore misleading to overgeneralise the lack of competitiveness of family firms. Cranton (1983, pp. 414–20) argues that family firms are as efficient as any other. Some of the most successful multinational companies in the world today, which include the Swiss Geigy and Maggi, Swedish Wicander, Rausing, Tetra Pak and Separator, Dutch Van Den Burgh and Jurgens and the Belgian Solvay are family firms. Thus, for Redding (1990), the vast majority of family firms, though small, have been the basis of the dynamic overseas Chinese capitalism, which has transformed the economies of much of Southeast and East Asia over the last three decades. In northern Nigeria, most of the successful businesses as earlier observed are family owned. Many of the leading firms in the world have proven that family ownership is not an obstacle to economic growth.[166] Some of the advantages of the family enterprises include the training of young family members and the development of a strong corporate culture that can be turned to competitive advantage.

In Kano, the major areas of commercial activities where family businesses are operated include Kurmi market, Singa, Katin kwari, and WAPA. The Kurmi market, is perhaps the oldest market in the city. It has served in the past as an entrepôt and a centre for long distance trade (Hill, 1977, p. 15). Nevertheless, the Singa, or Singer Market, which was built in 1838, is now the pivot of commercial activities. Many successful local merchant industrialists operate one or two commercial outlets at the Singa market. Because of its economically strategic location, most people perceive access to this market as a beginning to successful entrepreneurship. Ownership of a shop outlet in this overcrowded commercial niche therefore enhances an entrepreneur's rating in the business community. The large amounts of the wholesale items sold at Singa are essential goods for daily domestic usage such as sugar, flour, rice, salt, soaps, and milk. A permanent feature of the Singa market is the daily profusion of trailer vehicles unloading or awaiting delivery of goods for transport to other parts of northern Nigeria.

The diverse delivery roles of Singa merchants are very often more economically rewarding than manufacturing ventures. In contrast, the Kantin kwari is a textile market where indigenous and Levantine traders operate side by side. Entrepreneurial demand for foreign exchange to import foreign goods, especially textiles, has boosted activity in the currency market at WAPA. Most of the currency traffickers at WAPA are *Alhazai*[167] who possess tremendous wealth. Ironically, the traders and currency traffickers seldom pay taxes, and even when they do, the amount is negligible.

A reason for the low participation in manufacturing industry has been the relatively high failure rate of initial indigenous ventures in this field. Though there is a paucity of data on indigenous modern manufacturing activities in Kano, unofficial sources show that the rate of failure of indigenous manufacturing firms in the post-independence era has been very high.

A primary reason for the high failure rate of indigenous manufacturing firms in Kano is that most operate in a purely commercial fashion. Manufacturing activity otherwise thrives on continuity, and especially the smooth supply of raw materials, production and marketing of produced goods. The management discipline required for a manufacturing activity is often lacking in those experienced in commerce and trading activities. Sometimes, the persons put in key controlling positions lack the strategy, awareness and foresight necessary to maintain regular supply lines for their raw materials. A variety of factors including poor marketing and distribution networks, are also major handicaps to the local firms. Indigenous manufacturers frequently rely on the prices of goods produced by multinational firms which are already well established and have broad market base, to fix the prices of their products. Unlike multinational firms, most local firms lack the ability to absorb market shocks due to inadequate planning. These characteristics are typical of small family firms operated in many developing societies including Kano.

7 Resource Control Methods in the Plastic Firms

> Organisation implies control and control is an inevitable correlate of organisation (Dalton, 1971, p. 3).

In Chapter Six, I discussed the concept of profit as a motivating factor in business activity. I build on the profit analysis by showing that entrepreneurial and managerial resource control methods are central to the attainment of surpluses in the modern organisation.[168] Poorly managed resources not only lead to losses in the modern business organisation, they also inhibit its capacity for success. The subject of business success is examined in the Tenth Chapter of this study. Thus, entrepreneurial resource control strategy also has a direct bearing on the subject of rationality discussed in Chapter Three.

The relative scarcity of resources in modern organisations calls for managerial control of human and material resources. Resource controls therefore become the means through which entrepreneurs and managers 'assure that actual activities conform to planned activities' so that indications of any deviations are detected for corrective action to be taken (Daft, and Macintosh, 1984, pp. 43-66).[169] Pfeffer (1981) distinguishes between internal managerial and external environmental controls that impinge on every organisation, such as market conditions and macroeconomic and political policies.

Two contesting models are discernible in managerial resource control analysis: stringent and flexible controls. The former, which bears a resemblance to McGregor's (1960) 'theory X style of management' is based on the principle that the only way to get a task accomplished is through strict controls and direct supervision. A variant of this is the 'theory Y management style' which is based on the assumption of mutual trust and on the principle that employees are capable of providing a creditable performance even without close supervision (ibid.).

Although the stringent control method is largely considered old fashioned in the industrialised parts of the world, it is gaining ground among the managers of firms in developing countries. Relatively strict managerial controls were observed in the three plastic firms with varying degrees of emphasis. Kanter's

(1983) study of the 'forty-seven most progressive companies' in the United States of America reveals that managerial flexibility and employee empowerment are the most efficient ways of producing profitability, growth, and success. The findings of her study further show that stringent controls stifle creativity, profitability, and organisational success.

Societal values, beliefs, and conventions, sanction managerial decisions about appropriate control measures and the areas, and aspects of work that require stringent controls. Furthermore, the duration of specific control methods are value issues embedded in culture. Control measures are introduced in formal organisations for ensuring efficient combination of resources and proper reinvestment of profits. For Hechter (1987) a control measure is necessary for directing authority in any human activity that involves the combination of large-scale labour for securing 'harmonious working of the individual activities'.

Institutionalisation of control is necessary where authority is delegated to ensure the proper execution of tasks and the prevention of members from breaking rules with impunity. By institutionalisation, we refer to the processes by which obligations or actualities come to take on a rule-like status in social thought and action. An institution consists of a set of rules that provide guides to appropriate behaviour in a particular setting (Burns, and Dietz, 1992). Rules specify behaviour that is considered deviant and provide punishment for the deviant behaviour (John, and Perry, 1976). Institutional activities can further be classified as formal and informal. The formal activities include those activities prescribed by the organisation, which often become integral to the organisational structure. Informal activities on the other hand are those that develop spontaneously among workers in an organisation.

Two types of formal rules can be recognised within an institution; these are procedural and substantive rules. The former largely consists of the rules of the game, which deal with how certain routine activities on the shop-floor and relations on the production line should be conducted. Substantive rules, however, are managerial prerogatives that relate to decisions about wages, salaries, and allowances. Very often, informal rules emerge among workers that may threaten the general goals of management (Tichy, and Fombrum, 1979, pp. 923-965). However, informal activities often become a means of measuring employee motivation and discontent.[170] Whereas small groups can survive on informal controls, formal control is a prerequisite for the survival of large groups (Hechter, 1987). In a control situation, compliance from subordinates is necessary to ensure organisational harmony. Sometimes, strategic adjustments do occur among the rank and file of employees when they discover that specific rewards and punishments are contingent upon their performance (Leibenstein, 1975, pp. 580-606).

It is intriguing to note, however, that control methods are usually unpopular with employees. Drucker (1985) contends that the negative connotation associated with the word 'control' derives from the feeling that it threatens personal freedom and autonomy. This is because control denotes 'a tight hold by management on all things relevant to the operation of a firm' (Stoner, and Freeman, 1989, p. 555). Stringent controls and employee layoffs often become options or strategies for business proprietors in times of economic decline (Levine, *et al.* 1981). Since the rationale for introducing controls is to curb excess and waste, in effectively controlled institutions, the tendency for squandering by corrupt managers and supervisors is curbed. The relevance of studying control patterns in Nigerian factories stems from the fact that their failure is often attributed to poor managerial control methods.[171]

The introduction of formal controls in firms implies a power imbalance between those who design the rules, such as mangers, and those to whom the rules apply, such as employees. Power control, according to Biersteker (1978), is not limited to workers alone, but also to managers. Occasionally, proprietors of international or multinational firms introduce measures, or strip certain managers of their responsibilities to curb the excessive influence of indigenous managers over company policies. A common strategy employed by international or multinational firms, especially in Nigeria, is the selection of board members from across diverse ethnic groups so that disputes among them could limit or distract their ability to dominate foreign business associates (ibid.). These control tricks are not limited to foreign firms alone, but are used in indigenous firms as well. Most employers of labour in developing countries have come to accept the maxim that the extent to which they exercise control over their workers determines their ability to keep wages low.

Interaction between many individuals in a control situation makes communication an important instrument for the effective functioning of an enterprise (Sathe, 1983, pp. 31-48). In many Nigerian firms, including those covered by this study, the interaction of people from diverse ethnic backgrounds frequently leads to cross-cultural communication problems. Businesses in Nigeria, whether indigenous, international, or multinational, control people not only from token or bicultural backgrounds, but from multicultural backgrounds as well.[172] With over three hundred ethnic groups in Nigeria, many dialects that are mutually unintelligible are spoken in the country. Tension and antagonism often arise from intra-group misinterpretations, and misconceptions of messages and behaviours. It is necessary to mention that the tensions and cases of mistrust often created are usually the result of diverse cultural misinterpretations rather than of dislike or hatred.

Formal Controls and Business Performance

The enormous division of labour and diverse functions in the modern manufacturing firm implies that no single control device is perfect.[173] Multiple control measures with varying degrees of emphases were observed in the three plastic firms. Various departments in which controls were established include the accounting, production, and sales departments. For Webb *et al.* (1969) the only way to increase accuracy in industrial organisations is to use simultaneously in different departments 'multiple measuring procedures'. This does not imply that multiple control methods are without negative effects. Etzioni (1961) observes that simultaneous emphasis on two or more control methods may be counterproductive, as the measures often tend to neutralise each other. In the Chinese plastic firm, for instance, overemphasis on production controls led to neglect and poor supervision of the accounting department, which resulted in financial misappropriation by a Nigerian accountant.

The use and implementation of control measures has three different stages, which include pre-action, steering or cybernetic, and post-action stages. The pre-action stage is almost a preventive measure, thorough which planning is done on human and material resources before any action is taken. Financial resources are often carefully budgeted during the beginning of the year, month, or week before action is taken. In the steering or cybernetic stage, measures aim to detect errors and to steer away from them. The post-action stage involves taking a corrective measure after an error has been detected (Stoner, and Freeman, 1989). It is necessary to mention, however, that the fact these control strategies exist does not imply that managers in developing countries know about them or are willing to apply them.

Managerial attitudes regarding pre-action controls in both the indigenous Hausa and Lebanese firms were very poor and, sometimes, absent. One area in which this was noticed was the erratic power supply, with its attendant negative consequences on production. Operators of the two firms relied entirely on an external power supply from the Nigerian Electric Power Authority (NEPA). Despite the frequent interruptions on production, and accompanying reductions in daily production outputs, very little was done by the management of these firms to provide alternative sources of power supply. Occasionally, the workers in the indigenous Hausa and Lebanese firms had to be sent home because of electric power failure.

Intermittent interruptions of the production process in the two firms had several implications in terms of output. First, molten plastic material sets after cooling. This means that raw material during fabrication becomes waste and has to undergo a new recycling and manufacturing process. Evidence from the plastic factories has shown that sometimes after restoration of power supply, it took a considerable amount of time to get machines heated up to required temperatures for the specific manufacturing activity. Unlike the Hausa and Lebanese firms, the installation of two giant electric generators in the Chinese firm made a smooth supply of electric power and the conduct of production activities without interruptions. On a comparative note, the preventive or pre-action control measures in the Chinese firm were better than those in the Hausa and Lebanese firms.

In the manufacturing process, production is the key element to which profitability, growth and success are linked. In the industrial sector, the common tendency to measure performance against production capacity makes management control of production very relevant. Because of the whole range of activities that go into the production process, especially in manufacturing firms, production control sometimes defies precise definition. In this study, however, production control simply refers to managerial regulations on the quantity and quality of both input and output in the manufacturing process. Controls are established in the production process to maximise output, reduce waste, and to ensure that high-quality and not substandard goods are produced. The strengths and weaknesses of these measures depend upon managerial style.

A combination of observations and interview responses from factory employees revealed that the Chinese management employed the strictest control methods. Production controls in the Chinese firm were, in more mundane terms, highly autocratic and task-oriented. With the Lebanese plastic firm, the control measures were moderate, whereas in the indigenous Hausa firm, they were relaxed. Relaxation of controls in the indigenous Hausa firm, however, does not necessarily imply flexibility by its managers. Shop-floor or line supervision by Chinese management was particularly close. This was a measure adopted to prevent workers from shirking or indulging in unproductive behaviour during working hours. Occasional mistakes committed by workers on the line resulted in immediate threats of pay cuts. These measures had several implications for management-worker relations and the general working climate in the firm. Tension and a constant fear of dismissal among workers characterised the atmosphere in this firm.

As a strategy for ensuring higher productivity, the working hours in the Chinese were much longer than those in any other firm in Kano. While the normal number of working hours in the Hausa and Lebanese firms was eight hours, in the Chinese firm it was twelve hours. The daily working hours for the first shift in the Chinese firm normally began at 7.00 am and ended at 7.00 p.m. The workers in the Chinese factory, therefore, worked four extra hours than most of their colleagues in the other plastic factories.[174] The highly task-oriented Chinese firm was perhaps a reflection of the Confucian cultural work ethic. The eight naira, extra incentive given to workers as overtime allowance for four additional working hours was rather negligible, considering the amount of the work involved and the high rate of inflation in the country. It may seem that most of the workers in the Chinese firm complied with the extra working hours' rule for preserving their jobs, rather than for the additional income.

Unlike the indigenous Hausa and Lebanese firms, the Chinese firm occupied a monopolistic position as a pioneer plastic footwear company in Kano and probably northern Nigeria. The overwhelming demand for its products meant that, often, management had to step up production by compelling employees to report to work during the weekends. While conducting this study, four workers were dismissed in the Chinese factory for failing to show up during a weekend overtime assignment. This behaviour was contradictory to the provisions in the Nigerian labour law that makes work during the weekends voluntary.[175] Compared to the Chinese firm, the workload in both the Lebanese and indigenous firms seems fair.

A cursory look on the performance of business firms further reveals that financial resources are the pivots on which firms revolve. Improper financial controls often lead to bankruptcy and collapse of business enterprises. In most firms, financial and accounting controls are usually strict. This is because the accounting department, which deals directly with financial matters, is the most vulnerable area for improbity. Financial control refers to the methods used for monitoring cash inflow and outflow through good bookkeeping methods. This also includes budgeting, provision of good bookkeeping facilities and auditing (Vickery, 1973).[176] In the bookkeeping practice, the double entry method and periodic balancing of account books ensures the detection of errors, omissions and false entries. Accounting departments in Nigerian firms are frequently accused of the use of 'ghost names' in payment vouchers and of inflating contracts. In other instances, incorrect figures are given to managers and entrepreneurs, whose knowledge on accounting is limited. Wrong accounting figures and poor bookkeeping practices usually lead to managerial hunches and poor planning.

Bookkeeping practices of the Hausa firm were the poorest of the three plastic firms. The reason for this was probably that the accountant of this firm was not professionally trained. In spite of the poor bookkeeping practices in the indigenous firm, there was no recorded case of fraud in the accounts department. The local owner, who was seen by his subordinates as a 'very thrifty person', could attribute a reason for this to the direct supervision of all financial transactions of the firm. The bookkeeping practices in the Lebanese firm were better than the indigenous Hausa firm. Like the Hausa firm, there was no reported case of financial malpractice or any form of embezzlement in this firm. However, to ensure strict control of financial transactions, the Lebanese director and a Nigerian accountant of the firm jointly handled all accounting matters in the director's office. Paradoxically, the Chinese firm, which maintained the best bookkeeping practices and a qualified accountant, recorded a major financial fraud a year before this study was conducted. The huge financial losses suffered by the Chinese firm led to a large-scale dismissal of employees from the firm. Evidence from the Chinese firm suggests that good record-keeping practices alone are not enough guarantees for financial safety, some amount of honesty by business operators is also required.

Changing employee behaviour often becomes a necessity in any organisation where it proves to be unsatisfactory. On the other hand, its sustenance or maintenance may prove to be relevant where it is seen as satisfactory or positive. The term 'social control' is a broad concept that is subject to a variety of interpretations. In this study, social control is restricted to workers' behaviour in relation to punctuality, lateness, absenteeism, shirking, and labour unrest. The purpose of every control measure is to change human behaviour (Hofstede, 1978), or to prevent its change (Czarniawski-Joerges, 1987). Targeted persons in this regard are often employees under managerial control. Varieties of employee behaviour are dependent on upon their motivation. Employee behaviour often reflects the general working conditions and atmosphere under which workers find themselves. Different control strategies for motivating employees were observed in the three plastic firms.

For ensuring employee punctuality, the Chinese firm provided transport services for its employees. In addition, a health clinic was built within the factory-yard to care for injured or sick employees. Only extremely serious cases of injuries or ill health went to the general hospital for treatment. In contrast to the Chinese firm, both the Lebanese and Hausa factories had no health clinics. In view of this, employee health problems received treatment at the public hospitals and in private clinics at the expense of the firms. Health or medical incentives in all the factories were restricted to workers only. Employees' family members were not covered by the medical incentives. The Chinese plastic firm also provided a staff food canteen for its employees. Ironically, most of the local factory employees could not afford the expensive meals served in the factory canteen because of their poor wages. This development further widened the gap between the employees and management staff of the Chinese plastic firm. In the Hausa and Lebanese factories, the absence of factory food services meant that the workers relied on food sold by hawkers outside the factory gates. The food sold by hawkers, though more affordable, was often hastily and poorly prepared.

Nevertheless, the Chinese management was the most rigid in terms of the application of disciplinary measures. Lebanese and Hausa managers, who were less rigid than the Chinese, employed quite similar control strategies in their firms. Perhaps the reason for the similarity between Hausa and Lebanese management styles was probably that they shared a common Islamic culture. In both the Lebanese and the indigenous Hausa firms, employees were not subjected to extra working hours. Occasional weekend production activities created by the force majeure of market demands were voluntary. Factory workers were never compelled to report to duty as was the case in the Chinese factory. Attendance for extra working hours in the two firms was voluntary. Consequently, employee absenteeism did not carry stiff penalties and threats of managerial dismissal in the Lebanese and Hausa plastic firm. Cases of employee absenteeism were therefore more rampant in these factories than in the Chinese firm. As a control strategy for absenteeism, management in the Hausa plastic firm had to introduce a monthly financial bonus for regular attendance to work. An employee who was absent on three occasions in one month was unqualified for the bonus. This incentive had a positive influence on the workers in the indigenous Hausa factory.

It is pertinent to mention, however, that union activities were totally absent in both the Hausa and Lebanese firms. Levantine firms in Nigeria are notorious for the proscription of union activities in their firms. The reason for this is the desire of most Lebanese business proprietors to pay very low wages. In the indigenous Hausa firm, management staff had argued that there was no need for union activities since they were 'taking adequate care of employees' needs'. Most of the poorly educated workers in the Hausa firm did not reveal any keen interest or desire to have a union. This was probably due to previous disappointments with factory unions. Most of the employees complained about the collection of monthly dues by union leaders who failed to cater for their needs. Despite the lack of employee unions, the working environments in the Lebanese and Hausa firms were without the kind of tensions observed in the unionised Chinese factory.

The impact of control measures in a firm may be negative or positive depending on how they are employed and the extent of sensitivity they create on materials and among people. Concerning production controls, it has been explained that the measures introduced in the Chinese firm were the most stringent. Although the volume of production was generally higher in this firm than in the other factories, there were also, more cases of discontent in the Chinese firm than in the other plastic factories. The patrilineal attitude of the Hausa entrepreneur produced a relatively happier work force and encouraged employee lateness and nonchalant attitudes. With particular regard to financial controls, the study has shown that the direct financial supervision in both the Hausa and Lebanese firms minimised fraud. Managerial reliance on external auditors in the Chinese firm led to financial misappropriation. The evidence in this study shows that stringent control practices in the three plastic firms produced environments of pervasive distrust between employees and their managers in especially the Chinese and Lebanese firms. This perhaps suggests that flexible, controls which were lacking in all the firms, is a better alternative for generating trust and honesty in Nigerian firms.

Product-Based Trustworthiness

New ventures and products particularly need trust to capture the market. For a new product to attract public or consumer trust, it has to establish a track record. In particular, there must be a good value for the consumer's money. Products here refer to goods produced through mechanical or industrial process. Consumer preference of foreign goods in Nigeria, implies that product based trustworthiness for locally manufactured products is generally low. In spite of this, the presence of over 60 plastic companies in Kano suggests that consumer patronage of local plastic products is quite impressive.

Corporate reputation is very important in the establishment of business and product-based trustworthiness (Bromley, 1993, p. 1). Thus, corporate reputation, refers to a set of outstanding attributes ascribed to a firm such as quality of products, reliability of delivery, and fulfillment of agreements with customers. Reputation is relevant because it affects consumer choice regarding a specific firm. The quality of product is relevant for organisational or individual reputation because of the notion that 'a person is as good as his deeds'. In recent years, product-based trustworthiness is associated with good quality. The sustenance of high quality products by a firm is often the basis for attracting consumer trust. Quality sustenance is important because it reflects honesty and reliability. Product-based trustworthiness can be increased through durability as well regular and prompt delivery. Irregular delivery or the supply of non-durable products can be cause of loss of customers.

When consumers make purchases, they trust that the producer will be honest to provide good quality products. Product credibility can be lost if quality standards are not met. Quality variation is generally perceived as a sign of dishonesty and non-reliability. The ability to distinguish between good and bad quality products is as essential as being able to detect falsification (Dahl, 1998). Good quality products attract both trust and public patronage. In the manufacturing industry, quality control has become the medium, for ensuring the production of high quality goods. Production managers and line supervisors in the three factories were mostly involved in the quality control activities. Low quality products in the factories were separated from the rest of the products and returned to recycling section for crushing. Referring to the market in which Nigerian products are sold, Schatz observes that:

> The market for the output of these enterprises is relaxed about product uniformity, product specifications, and regularity of production schedules. It is therefore possible for a firm to get by despite serious deficiencies in production, delivery, services, and sales promotion (Schatz, 1977, p. 78).

Evidence from the three plastic factories suggests that small plastic manufactures often compare their products with those of competitors to find out the quality of their products. In addition, they relied on customer complaints to improve on the quality of their goods. Although the criteria for providing international awards are not known, it is interesting though to mention that three plastic factories covered by this study, had once in their business carriers won the prestigious Geneva International Awards for 'good quality products'. The awards which where conspicuous displayed in the three factories were used as a medium for convincing all customers who visited the factories.

Trust is more likely to be extended to an organisation when it earns a reputation in the market for following norms of equity. In view of this, the plastic products from the Chinese firm attracted more trustworthiness in the market than the others. The reasons could be attributable to fact that the Chinese conformed to higher quality standards of production and applied more uniform production standards than the rest.

A general assessment of the three plastic firms further reveals that none of them had a quality assurance department. The most common strategy for quality control was occasional manual inspection of products by supervisors and managers. Entrepreneurial quality adjustments were contingent upon the comments received about their products from customers. The prescribed standards in the plastic industry for continuous quality control at each specific work cycle were seldom met in the plastic firms.[177] A Lebanese manager observed that quality control measures in the plastic industry were expensive and only rich multinational firms could afford them. The lack of enforcement of strict quality controls by the Standards Organisation of Nigeria (SON), in the manufacturing sector has meant that the qualities of products, including those from the plastic industry, are sometimes low. Manufacturers in the Nigerian plastic industry sometimes employed deceptive strategies to boost their sales. A common strategy was the choice of colours for products. An indigenous plant production manager observed that:

> A common method we use to boost the sales of our products is in the selection of the colours used in designing them. In this regard, the masterbatch concentrates, the chemicals used in the plastic industry for coloration, are indispensable. Generally, customers tend to believe that darker products are more durable. Even when the products are of the same material and quality, colour sometimes becomes the means through which consumers differentiate reliable products. These peculiarities, which are unknown to some of our competitors, are our strategies for boosting sales.

This specific example shows that cultural conceptions about colour often influenced customers' preferences for industrially manufactured products, especially in the plastic industry. This preference was brought to bear on the production strategies of local manufacturers of plastic products.

A common strategy for capturing or gaining access into a market by smaller firms, is in the pricing of a new product. For a new product to compete favourably with well-established ones in a particular market, it has to sell at a lower price. Low pricing is therefore important for a new product. New products in a developing society, often encounter suspicion and therefore require a tremendous amount of publicity to capture the market or boost trustworthiness. New venture operators according to Macmillan (1983, p. 8) must learn to anticipate resistance and reluctance, and develop political skills to neutralise and overcome them. For new products to capture the market, especially in northern Nigeria, consumers must be educated on whether specific products do not threaten their cultural or religious values.

A typical example of the resistance to a new product to the market was the introduction of foreign soft drinks in the northern Nigerian market in the early 1950s. The introduction of these new brands of soft drinks, including Coca-Cola, was met public resistance from the predominant Muslim population of northern Nigeria. Hitherto, many soft drinks were suspected in the Muslim communities as alcoholic products. Bottling companies deployed series of strategies through advertisements and the invitation of indigenous shareholders. With the passing of time, soft drinks have captured the entire markets of northern Nigeria. To enhance public confidence in their products, manufacturers of soft drinks in the region often inscribe on their bottles 'this is not an alcoholic product' or *ba giya bane*. Sometimes containers in which purified drinking water are bottled, have inscriptions informing consumers, 'this is not alcoholic water' *wannan ba ruwan giya bane*.

Manufacturers in Kano, often use advertising strategies to draw consumer awareness to new products. Such adverts often involve the use of the media such as radio, television and posters. Sometimes marketing personnel, sales staff and merchandisers visit communities, markets and business outlets to sell new products. Such advertising strategies are common among multinational firms, which have both materials and employees to embark upon such exercises. It may therefore seem that to ensure product trustworthiness, manufacturers have to produce good quality products, to draw consumer attention. Furthermore, manufacturers have to conform to the economic demands of prompt delivery as a way of building enduring reputation.

Union Problems and Employee Discontent

With a population that is double the size of Francophone West Africa, and an active labour force of over fifty million people, Nigeria has the largest labour force in sub-Saharan Africa. Ironically, less than 20 percent of this labour force is unionised or belongs to any employee associations. The preponderance of weak unions is not unique to Nigeria, but is a common feature in most developing countries where democratic ideals are not well entrenched. Trade unions are both a representation of vital interests in a pluralistic society as well as an expression of freedom in a democratic setting (Sonubi, 1987:76). Trade unions tend to assume the political, economic, social, and cultural characteristics of the environment in which they operate. It has been earlier explained that Nigeria is a mixed economy. In a mixed economy, both the state and private individuals employ workers. This implies that workers are found in both public and private sector unions (Otobo, 1987, p. 47). The sector in which they find themselves and the way the unions are operated condition employee attitudes towards their associations.

Trade unions are interested in four major issues, which include, the price of labour, working time, equality or fair play, and power. Very often, the demands of workers in these areas are antithetical to the interests of their employers. It has been observed that Nigerian workers are interested in the price of labour, fair play, and improved conditions of work.[178] The quest for power is secondary to most Nigerian workers. The prevalence of indigenous family businesses, large companies, parastatals, and subsidiaries of multinational firms means that arriving at compromises across comparable firms of unequal economic strengths are major problem employers have to face.

Since their formative years, Nigerian trade unions have been bedevilled with many problems that have led to significant structural changes.[179] The federal government in 1976 undertook the most restructuring exercise. The untidy assemblage of 721 weak house or plant unions was reduced to 70. Of this figure, 42 were designated as industrial unions for junior workers. In a 1976 restructuring exercise, 15 unions were designated as senior staff associations, four as professional unions and nine employer associations. All the unions operate under the umbrella of one national union known as the Nigerian

Labour Congress (Otobo, 1987, pp. 45-71). Presumably, Nigerian workers in the plastic industry are *bona fide* members of the national Footwear, Leather, Rubber Products Workers Union of Nigeria.[180] In real practice, however, very few plastic firms are actively involved in union activities or have any affiliations with the national union. Employers of workers in the plastic industry belong to a similar sister union known as the Footwear, Leather and Rubber Products Senior Staff Association. However, it should be mentioned that few small or medium-sized firms in the plastic industry are unionised.

All employees in the Chinese factory belonged to a union. Nevertheless, the tight controls introduced by the Chinese operators' of the plastic firm made the worker's union very weak. While weak unions or their nonexistence may give rise to the marginalisation and exploitation of factory workers, it does not follow that the evolution of strong unions will pave way for industrial harmony (Fashoyin, 1980, p. 34). It therefore became a paradox that the unionised workers in the Chinese factory were the most antagonised when compared to their non-unionised counterparts in the Hausa and Lebanese firms. In the Chinese plastic firm, the prominence of non-indigenous workers in union activities created intense rivalry between them and indigenous Hausa and Fulani employees. Management and union leaders sometimes exploited the ethnic heterogeneity for personal gain. Religious divisions further exacerbated intra-employee competition for the union leadership, with the predominantly Christian workers from the southern and middle-belt areas presenting a common front to compete with their Muslim counterparts. The inability of the union leadership to reach a successful agreement with Chinese management on eight normal working hours resulted in employee lack of confidence in the union. Consequently, most of the indigenous Hausa and Fulani factory employees were largely inactive in the voluntary monthly union meetings.

Several factors combine to produce weak unions in Nigerian firms. First, the repressive and militant approaches by various Nigerian administrations since the colonial era, when unions were a major pressure group at the forefront of the independence struggle, has not altered in post-independence period. Because of the rather poor rapport between government and Nigerian unions, nearly every policy dealing with union activities has attracted some amount of suspicion. The restructuring of unions by the government in 1976 and in 1987, was seen by some observers of the Nigerian industrial relations system as a deliberate ploy to foist 'docile and effective unions over the country' (Sonubi, 1987, p. 80). Structurally, the hierarchical arrangement of unions along national, industrial, and company lines and the poor coordination between them accounted for their weakness. In Nigeria, government is the major employer of labour, and employers of labour in the private sector have sometimes wrongly emulated some of its draconian policies and deplorable attitudes towards unions.

Second, the very idea of workers' unions appears alien to most African cultures. The tacit notion of a trade union, or *kungiyar kwadago* lies in the recognition of differences and diversity of interests. Culturally, the traditional African setup with its closely-knit family system, generally perceives the ideas which trade unions represent as an aberration. Cultural values and norms tend to place emphasis on corporate harmony and unity rather than diversity. The ideas of diversity that trade unions appear to represent, are often seen as strange and unhealthy for business operation. In most indigenous cultures, diversity of interests in a group may be interpreted as failure of leadership, which many indigenous entrepreneurs strongly try to resist. Sometimes unions are deliberately outlawed in Nigerian firms not only as a repressive measure, but also as an attempt to prove to the outside public that things are running smoothly to an extent that trade unions have become unnecessary. On the union question, a local personnel manager in a plastic factory made the following observations:

> We do not need unions here. Unions may be necessary where workers and management have differences to resolve. Here, there are no differences . . . only common interests. The workers in this factory are all happy because we see ourselves as a family. Why then, must we spend time creating unnecessary unions?

A recurrent response from Hausa and Lebanese managers regarding unions has been an overwhelming rejection and branding of unions as 'unnecessary'. Hausa and Lebanese business proprietors who together constituted 55 percent, were of the opinion that unions were 'problematic' and 'full of unrealistic demands' often incommensurate with productive effort and profits. This view, however, produced a dialectic between managerial and employee values in the plastic firm. It also contradicts the modern capitalist notion of equal representation in organisations involving diverse group interests. Although three-quarters of the Chinese managers favoured unions, their tough attitudes towards the employee union posed a contradictory picture. The evidence in this study suggests that entrepreneurial suppression of unions in the plastic sector was based on no other reason than their desire to pay low wages. The truism, according to Karl Marx, is that in virtually every struggle between the capitalist and worker, the former inevitably wins because 'the capitalist can live longer without the worker than the worker can live without him'.[181]

Thirdly, the poor attitudes of workers themselves towards unions were major hindrances to their survival. Inadequate information and the low level of education of most of the workers, especially in the plastic industry, were factors responsible for low employee participation in union activities. In the urban Kano environment Andrae and Beckman (1998, p. 25), observe that the 'informal modes of recruitment and clientelistic relations between labour and management constrained the scope for unionisation'. Paradoxically, the strength of unions lies in their organisation and the active participation of members who are aware of their shortcomings (Sonubi, 1987, p. 86). Of the 166 employees interviewed, 91 of them, representing approximately 55 percent, did not belong to any unions. Employee lack of interest in union activities was because most unions were too weak to make any meaningful impact on their demands for wage increases or improved working conditions.

The need to replenish dwindling funds and to ensure adequate and regular sources of income for Nigerian trade unions has led to the introduction of the compulsory checkoff system in most organisations. Increases in union funds in recent years, coupled with inadequate controls, have accelerated financial mismanagement and subversion within the rank and file of union leadership (ibid.). The growing dissent among the lower ranks of union workers in most factories is the product of feelings that the monthly compulsory deductions from their meagre incomes are misappropriated. The union dues, which are compulsory in unionised factories, are usually token fees that do not usually exceed 4 percent of an employee's monthly earnings. Nevertheless, because of the difficult economic situation in the country, such compulsory deductions from workers' earnings, no matter how little, can be the source of discontent and anger among salaried employees.

Conflict Resolution Procedures

The peaceful coexistence between entrepreneurs and those they employ depends upon a variety of factors. These include the level of wages paid to workers in relation to those paid in comparable firms, working conditions, and entrepreneurial attitudes regarding freedom of movement and expression. Apart from the low salaries and poor working conditions, cross-cultural misunderstandings are often the source of tensions and discontent among factory workers. The prevalence of peace in any factory is a major factor for high productivity. Employee grievances in factories represent their dissatisfaction with the conditions of employment.

Several grievances were recorded among employees in the three plastic factories. However, grievances regarding conditions of work were most severe in the Chinese factory. Before this study took place, the Chinese plastic firm had experienced several violent industrial strikes in 1989, 1991, and 1993 often resulting in the use of police and industrial courts to settle disputes. In all three instances, poor wages and cross-cultural misunderstandings were the responsible factors. When confronted with the question on conflict resolution, the members of the Chinese management showed a unanimous predilection for police force in resolving industrial disputes. Despite their experience of two violent strikes in 1987 and in 1994, the Lebanese management was equally divided on the issue of use of coercion to resolve industrial conflicts. Most Hausa entrepreneurs and managers, representing 61 percent, were against the use of police and showed a preference for informal conflict resolution methods.

In Nigeria, the resolution of industrial disputes is governed by Trade Disputes Amendments Decrees of 1988, numbers two and thirty-nine.[182] The decrees set out stages through which disputes can be resolved. In the industrial sector, the stages include the plant, company, and industrial levels. Disputes or grievances at the industrial level are resolved through Industrial Arbitration Panels (IAP) and National Industrial Courts (NIC). Under the new amendments, which came into effect on 27 October, 1988, the two industrial courts have the power to enforce industrial awards. However, the evidence in Nigeria has shown that the functions of legal institutions are more theoretical than practical. Because of the cumbersome nature of the legal system, many Nigerian and Levantine industrialists often prefer informal dispute settling procedures over litigation.

The clientelistic recruitment patterns imply that conflict resolution procedures in the three plastic firms were largely informal. The basic advantage of the informal approach in conflict resolution is flexibility. It allows disputants to resolve their differences amicably without engaging the services of a third party. Moreover, the informal, and verbal conflict resolution procedures are very useful where most workers do not have formal education. Consequently, the methods applied in resolving conflict in many Nigerian firms depend on the extent of network relations, nature of the grievance and the parties involved. In spite of this, most of the informal procedures for resolving conflicts are flawed because of their pro-management attitudes. Much of the evidence in the three plastic firms suggests that any enduring peace in the factory must recognise workers as partners in progress rather than saboteurs.

8 The Informal Sector Economic Activities

> A sector to which entry by new enterprises is comparatively easy; Enterprises in this sector rely on indigenous resources and are family owned; they operate on a small scale, in unregulated and competitive markets and use labour-intensive and adapted technology; their workers have skills acquired outside the formal school system (ILO, 1972).

This Chapter, reviews activities in the dual economy of northern Nigeria by attempting to show the juxtaposition between the informal and formal business sectors. Of particular interest to the analysis is the interplay between precapitalist and capitalist modes of production in these two sectors of the economy. The distinct modes of production significantly influence different forms of rationalisation discussed in Chapter Two. The second section of the Chapter, looks at institutionalised migration as an economic factor, and its role in the supply urban surplus labour, a key actor in the dual economy. Cultural conceptions and attitudes about time, and the activities of expatriate and local business groups in the dual economy are central to the discussion in the final section of this Chapter.

The most common terms for describing African economies in recent times are the informal and formal sectors.[183] These concepts, which were popularised by Keith Hart (1971) perceive the developing or underdeveloped economies as 'dualist'. Two sectors are recognised in this economy. The dominant capitalist sector is known as the 'formal sector', while the sector that operates outside it is dubbed the 'informal sector'. The informal sector is synonymous with the 'marginal sector' analysis of 'mainstream capitalism' used by Latin American theorists (Worsley, 1984). Significantly, the informal sector reflects a unique characteristic of free enterprise within the indigenous African cultural fabric. The form or nature of the informal sector reflects the peculiar characteristics of the cultural values and norms of the particular environment.

Two schools of thought are discernible in the informal and formal sector analysis. These include the 'reformist school', championed by the International Labour Organisation (ILO), and the Marxist school, strongly favoured by radical scholars from the developing world. The reformist school perceives the informal sector as having a tremendous potential capable of propelling economic growth in the less developed countries. A main thrust and focus of the Marxist school are the effects of the structural relationship between the informal and formal sectors and the capitalist world economy. There are, however, two divergent positions in this school of thought. The first position perceives the informal sector as consisting of a particular 'marginalised people' (Obregon, 1974). The second Marxist position, known as the 'petty commodity production school', considers the informal sector as subordinated to the formal sector through direct links (Moser, 1976; and Tokman, 1978).[184] A major weakness of the 'petty commodity' position is its failure to make distinctions among the various activities in the informal sector (Natrass, 1984).

The International Labour Organisation provides a codified distinction between the formal and informal sector labour in quite an illustrative manner. According to this codification, the informal sector can be distinguished from the formal one by easy entry. While entry is easy in the informal sector, it is difficult in the formal sector because of the requirement of special skills and tremendous amounts of capital. Second, while the informal sector relies on indigenous resources, the formal sector relies on imports. Third, the informal sector is characterised by individual and family ownership of the business enterprise, while the formal sector is often a corporate body. Fourth, the informal sector is labour-intensive and makes use of adapted technology, while the formal sector is capital-intensive and uses imported technology. Fifth, whereas skills used in the informal sector are found outside the formal school system, in the formal sector, they are obtained in formal schools and sometimes even abroad. Lastly, while the market in the informal sector is unregulated and competitive, the formal sector is not only regulated but also protected by tariffs, import, and exchange licenses.

Distinctions between the informal and formal sector are sometimes blurred because the two concepts are not mutually exclusive, but often overlap. Worsley's (1984) contention that the informal sector is 'a greatly reified concept' is most appropriate. In developing economies, the complementary roles of the formal and informal sectors make it difficult to distinguish between the two because of skills. As such, they are obviously not poles at the end of the continuum, as they may be portrayed. Furthermore, the informal sector is not just a countryside phenomenon, because it thrives even in the heart of the cities of developing countries. Usually, it serves as a leisure-time activity for employees in the formal sector trying to supplement meagre incomes. Thus, 'in its broadest definition the informal sector defies geographical specificity' since it covers a variety of activities virtually everywhere (Lubeck, 1986, p. 91).

An estimated share of the urban labour force in selected African countries reveals that the bulk of the urban labour force in Nigeria, Ghana, and Senegal is in the informal sector. Whereas Ghana topped the list with a 60 percent share of its urban labour force in this sector, Nigeria and Senegal followed closely with 50 percent each.[185] The preponderance of unemployed and under-employed labour in most African cities, combined with the quest by the lumpenproletariat[186] to survive under severe economic conditions, further boost the activities in the informal sector in most urban centres. In recent years, mass employee retrenchments in the public and formal sectors of the economy keep the informal sector swelling at a tremendous pace.

Because of the Islamic influence, 'open' informal sector activity in northern Nigeria is a predominantly male domain. Not surprisingly, only one out of the 44 business entrepreneurs and managers interviewed was female. In other parts of the country, the picture is different. In the eastern and western parts of Nigeria, most women are as actively engaged in open business activities as their male counterparts. These differences are perhaps attributable to the cultural values, and the impact of historical past of the people of these areas. Despite these differences, the secluded Muslim women of Kano continue to play significant roles in economic development through household craft occupations and by conducting petty trading through child hawkers. The relevance of the informal sector to the formal private sector is because the two economic sectors are inseparable. An examination of the informal sector further explains the problems of poverty and the behaviour of employees in many developing countries.

Gano's (1983) survey of the urban informal sector of Kano recorded over 10,000 hawkers engaged in different economic activities. This figure has been criticised by Mortimore (1989) as incomplete and an underestimation of the true position. Estimates from unofficial sources suggest that there are currently over 25,000 street hawkers of different categories in the Kano metropolis. Official records from the federal office of the Corporate Affairs Commission (CAC) in June, 1997, showed that there were over 70,000 officially registered businesses in the informal sector of Kano. This figure, which represents less than half the total number of businesses in Kano, does not include petty traders and hawkers.

In spite of the arguments for and against the informal sector, it is easily distinguished by the provision of services for immediate cash or rewards. Employment in this sector does not require strict rules or regulations for entry. The decision to commence business operation usually depends only on the individual. The fixing of prices in the informal sector is of a bizarre nature. Prices of goods and services in this sector are seldom fixed, but left open to the customer or client's ability to haggle and bargain. Thus, for every good or service, there are two prices: the real price and price quoted at the beginning of any bargain. The real price is arrived at after a prolonged bargaining session. Consequently, the same commodity is sold at different prices to family members, friends, and foreigners (Dia, 1992, p. 173). Business attitudes in the informal sector are often consonant with the precapitalist economic behaviour. Arbitrariness in price fixing for favoured customers sometimes ruins microbusinesses in the informal sector.

For Phillipe Hugon (1980), the informal sector may be classified into four categories. These include, service, small manufacturing, micro-commerce and transport sectors. Those in the service sector include, cart pushers, barbers, toe and finger nail clippers, radio, bicycle, motor bike and vehicle repairers. The small manufacturing sector according to Hugon include, cobblers, tailors, utensil makers, designers of metal window frames, potters and so on. The micro-commerce and transport sector operators in the informal sector include, peddlers of secondhand clothing, food hawkers, and taxi drivers. At the lower echelons of the informal sector are garbage collectors.

Compared to the formal or industrial sectors, the informal sector is characterised by the absolute freedom of the people who operate within it. For example, the number of hours to be spent at work and the time to report in is left to individual prerogative. The inhibition to participation in the informal sector is that incomes are unstable and sometimes below the national minimum wage. Nevertheless, the informal sector is a highly competitive sector. No sooner is an economic activity booming than nearly all others follow the same line of activity. This is often done without regard to the market potential of a particular product or service in the area. This scenario often leads to the stagnation and failure of indigenous businesses. Some economic activities in the informal sector involve seasonal and perishable consumer items, which cannot be relied upon for continuity. This tacitly implies that, to survive, sellers of seasonal food items do not rely on a single economic activity. The meagre incomes and job insecurity in the informal sector increasingly compel most people to seek jobs in the organised industrial sector.

Despite these handicaps, most successful business industrialists in Kano have previously operated in the informal sector. There is therefore enormous potential in this sector for economic transformation. Writing on the informal sector in West Africa, Demol *et al.* (1982, pp. 373–374) assert that this 'sector, without obstructing the modern formal sector, performs an important training function and income generating base for a vast majority of those engaged in it'.

As in most developing societies, there is a strong juxtaposition between the informal and formal economic sectors. For most workers in the formal economic sector, the informal sector is almost a training-ground where skills are acquired for the formal industrial sector. Most professionals and artisans, like electricians, carpenters, masons, painters, and drivers, usually find their skills in the informal sector before joining the formal sector. Similarly, it is a common trend for employees in the formal sector to switch over to the informal sector to establish small enterprises after getting their skills in the formal industrial sector. The informal sector has become a pastime activity for employees in the formal sector striving to supplement meagre incomes provided by wage employment. In the urban Kano society, it is common to find industrial and government employees working as taxi drivers after official working hours. The informal sector is undoubtedly a 'reservoir of dynamic entrepreneurs' (Ikiara, 1994). A World Bank report (1989, p. 135) describes the informal sector as 'a striking illustration of the strengths and weaknesses of business enterprises in Africa. Although unregulated and largely unrecorded, its activities comprise the most accessible and competitive economies'.

There is an overwhelming amount of unemployed or underemployed labour in the informal sector. In many Nigerian and Lebanese factories, unskilled employees from the informal sector are treated as casual workers. Casual or surrogate labour, as the name implies, does not make up a permanent part of the factory labour force, but is usually considered temporary. The services they offer to their employers are assessed and remunerated on a daily or weekly basis. During certain times of the year, especially in the dry season, large crowds of casual labour at the factory gates of Kano are very common. The excess supply of men in the Nigerian labour market makes the conditions of job seekers precarious. Adam Smith (1981; [1776, 1989]) observes that in any system where the supply of men greatly exceeds demand, a 'section of them sinks into beggary or starvation'. The reluctance of employers of labour to integrate casual labour into the permanent labour force often stems from the low level of education of most workers. Such employees are often not educated enough to know what forms employee rights even in the casual labour position. Because of the poor treatment of labour in the factories, the ultimate goal of most employees is to return to the informal sector as soon as they acquired sufficient capital capable of starting business (Peace, 1979; and Lubeck, 1986).

Some problems associated with the informal sector are usually management related. First, the methods of fixing prices are often very haphazard and do not follow any defined pattern. Second, most of the businesses suffer from lack of proper forecasting, coordination, and planning. Third, operators of businesses in this sector seldom attach importance to the 'economic time' (Demol, *et al.* 1982). Nevertheless, the informal sector offers many advantages, which include creation of jobs for unemployed people who would have been considered unskilled and unsuited for the formal sector. Furthermore, there is a stimulation of both economic and skill development, and the provision of assistance in income distribution. Consequently, the informal sector acts as a 'shock absorber' to the chronic unemployment problems confronting most developing countries. The informal sector therefore does not imply stagnation. Contrarily, it portrays a considerable amount of dynamism, which, if encouraged, can transform into a national modern sector, still firmly anchored on the locally distinctive social and cultural fabric (ibid.).

In contrast to the labour in the informal sector, employees of the formal sector operate in a structured system and are often very skilled or highly educated. Formal sector employees often conduct their affairs strictly under laid down rules and regulations which economic activities in the informal sector lack. Operators in the formal sector include entrepreneurs, business people, managers, engineers, accountants, cashiers, nurses and professional men and women. The services offered in this sector are not for immediate cash rewards. Compensation for employee services may be in monthly salaries, wages, and allowances.

It is, however, an over-generalisation to classify all labour in the formal sector as skilled or educated. Suffice to say that many people in the Nigerian formal sector are without any genuine qualifications. Apart from the steady monthly salaries that employees in the formal industrial sector enjoy, the living standards of most employees in this category are not much different from those of self-engaged persons in the informal sector. Despite the incredibly poor wages and frequent employee retrenchments, factory employment in the formal sector provides an economic refuge for the unemployed struggling with the vicissitudes of daily life.

Institutionalised Migration as an Economic Factor

Migration from rural to urban centres in Nigeria often provides the escape route from rural poverty. It is also the means for starting a small commercial activity in the informal sector or a factory employment in the formal urban economy. Not surprisingly, most of the successful business people and low-income factory workers have a rural background. Apart from the significance of migration to labour trends, some academics consider it a passé in contemporary discourse. Ubeku's (1983, pp. 53–54) study particularly portrays labour migration as no longer constituting problem for Nigerian firms. In contrast, the studies by Hill (1977), and Peace (1979), in Kano and the Lagos metropolis, have shown that migration is a key issue through which the problems of poverty and the behaviour of wage employees in Nigeria can be properly understood.[187]

Table 8.1. Projected Growth of Labour Force 1960–2000

	Average annual percentage growth rate			
	1960–1970	1970–1980	1980–1990	1990–2000
Developing Countries	1.8	2.2	2.2	2.1
Developed Countries	1.2	1.2	0.7	0.5
Asia and Pacific	2.4	2.6	2.3	2.0
Latin America	2.4	2.7	3.0	2.7
Middle East and North Africa	1.9	2.6	2.9	2.2
Sub-Saharan Africa	2.2	2.2	2.5	2.6

Source: World Development Report 1979.

The recurrence of the migration theme in contemporary development discourses not only emphasises its reality, but also shows the relevance of an adequate and steady supply of labour to enhancing sustainable economic growth.[188] As a catalyst to cultural change, a look at migration provides relevant insights to the social and religious dynamics in contemporary Kano. This scenario, however, has several concomitant ramifications for economic change, some of which are examined in this chapter. Since most Nigerian, and indeed African migrants move from rural to urban areas, their influx into cities has a positive correlation with business expansion. In spite of this, the evidence suggests a negative relationship between population growth and economic development.

Nigeria's population of 110 million people according to 1998 estimates makes it the most populated country on the African continent. With a birth rate of 42.58 per 1,000 and an annual growth rate is 3.05 percent, Nigeria ranks among the fastest growing nations of the world. In comparison to many African countries in sub-Saharan Africa, Nigeria's population is very heterogeneous. Rapid population growth in Nigeria has become 'associated with virtually all problems' (Isiugo-Abanihe, 1994, pp. 66). Rapid population increases have resulted in a meteoric 10 percent annual expansion of major cities such as Lagos, Kano, Kaduna, Ibadan, and Enugu. This has led to a rise in urban proletariat, and an increasing wave of crime.[189]

Today there is a growing wave of cynicism among demographers and economic planners that, although the Malthusian catastrophe may have been averted in parts of the industrialised world, the same assumptions cannot be made about Africa. High population pressures and the disproportionate effects of political instability are seen as factors responsible for large-scale human movements to major cities in sub-Saharan Africa.

The projected annual growth rate of the labour force in most developing regions including Latin America, the Middle East, and Africa as shown in table 8.1 is quite significant. While the annual growth rate of the labour force in the developed areas of the world is projected to decline to a mere 0.5 percent by the year 2000, in the developing regions, a rate of 2.1 percent emerges. In sub-Saharan Africa and Latin America the projected annual growth rates of 2.6 and 2.7 percent by the end of this century, compared to annual food production, is worrisome. Paradoxically, the large growth of the labour force in sub-Saharan Africa, and especially Nigeria, does not necessarily imply that there is a steady supply of skilled labour in the organised private sector. Most of the labour-force in these countries is unskilled. Consequently, only a small fraction of migrant labour in the urban areas of developing countries can be gainfully employed.

Labour migration in this text refers to the transfer of labour from a place of domicile to a destination of perceived opportunity or employment. Such labour remains migrant because 'no matter how long he/she stays, . . . The village remains home with which continuous contact is maintained' (Yesufu, 1962). The urge to migrate to a new environment is often boosted by a perception of better economic prospects in the place of destination compared to the place of origin. Two forms of labour migration have been recognised in northern Nigeria. These are the seasonal circulatory and the more permanent type of migration. In the former, the migrant is constantly in touch with home.[190] The most common reason for this is that very often, spouses are left behind in the villages. There is also a fear of losing land to false claimants in prolonged absence or when 'residence in the village is given up'. Hill (1982, p. 269), observes that this behaviour shows a link between migration and the maintenance of property rights in the countryside.[191] In their study of the nature of human settlement in Kano zone, Mortimore and Wilson (1965, pp. 35–7), caution that since there are various types and reasons for migration, it will be inappropriate to describe migration on strictly economic grounds.

Although economic, ecological, social, and political factors may compel a person to migrate, the decision is ultimately a subjective one. The subjectiveness of the decision involves values and beliefs about why movement is necessary, what destination is appropriate, and what duration is ideal for a particular destination. Thus, culture is in the forefront of migration analysis. The economic implications of migration should therefore be seen from within a cultural realm. Krokfors (1995, pp. 54–64) observes that 'cultural factors have to be taken much more into consideration in migration studies than has been done earlier'. Apart from economic motivations, a combination of factors relating to religion, kinship, clan, ethnicity, social status, and access to 'state or natural resources' may help to make people prone to migration (ibid.). Consequently, the circumstances leading to migration, whether voluntary or involuntary,[192] local or international, are many and complicated.

Perhaps nowhere in Nigeria is labour migration as noticeable as it is in the north. Labour migration in northern Nigeria is environmentally or ecologically induced. A combination of a short rainy season and a prolonged dry season compels self-employed labour to become idle for the greater part of the year.[193] Migration from the countryside to the city often becomes the escape route from unemployment and prolonged idleness caused by environmentally induced poverty. In recent years, migration has become an established strategy for augmenting incomes in Hausaland and neighbouring areas as far as the Niger and Chad Republics (Mortimore, 1989; and Hill, 1972). The link between season and migration is unequivocally expressed in the Hausa term for migrants or *Yan cin rani*, which literally translates as 'the men who spend the dry season'.

The Hausa definition of migration is linked to the activities in the agricultural sector. The movement of peasants during the dry season, from rural parts of northern Nigeria to urban centres, is perhaps a reflection of the fact that the lands in the northern part of the country are no longer agriculturally sustainable throughout the year. The recurrence of this seasonal phenomenon makes most migrants habitual seasonal commuters from rural areas to the cities (Yesufu, 1962, pp. 116–124). These culturally peculiar types of movements have various local Hausa qualifications.[194] In Hausaland, men migrate more frequently than their female counterparts. This is partly rooted in the Islamic culture, which restricts the movements of adult females. Most Hausa women who migrate only do so to join husbands who are already settled in other places (Hollos, 1991, pp. 852–870).

Female economic activities are largely restricted in the informal sector. The economic transactions of women in purdah, or *kulle,* are frequently conducted through a variety of intermediaries including spouses and child hawkers who shuttle among compounds in the cities (Smith, 1962). Pittin (1984) observes that some of the few Hausa women who migrate to the cities without their spouses are single, divorced, or widows who exercise tremendous initiatives in their abilities 'to parlay their earnings into considerable wealth through shrewd investment, careful planning and a great deal of hard work'.

Rural-urban drift can be seen within the context of rural surplus labour in developing countries (Ranis and Fei, 1961). Urban conditions have become avenues for attracting unskilled labour from low-income jobs to higher or at least steady paying jobs in the formal industrial or public sectors. Todaro (1969) observes that, apart from the expectation of earning higher wages in urban areas, the probability of obtaining a modern sector job within a year or two in the city influences the decision to migrate (1969, p. 140). Migration can therefore be seen as a function of rural-urban wage differentials and the probability of getting a job (Fadayomi, 1988). Lubeck's (1983) survey of 3,031 employees in Kano reveals that 54 percent of them migrated to the city to seek new economic opportunities, whereas 22 percent came to seek Western education. Significantly, 10 percent of the respondents in the survey had migrated to Kano because of personal relationships or family ties. The harsh conditions of rural life and social networks created during the years of residence in the city, coupled with greater opportunities for trade and economic independence, sometimes act as barriers for the return of migrants to the countryside.[195]

The directions of the flow of migrant labour include, rural-rural, rural-urban, internal and international (Amin, 1995, pp. 29–40). While some migrants go directly to their destinations, others use step-migration, moving from smaller settlements to progressively larger ones in a hierarchical order. Some migrants from the northern parts of the country engage in what is known as 'multiple migration' (Peil and Sada, 1984, p. 121). In the multiple method, the migrants search for opportunities in both rural and urban areas, although this is less common than migrating directly to the cities.

Unlike their southern Nigerian counterparts, most dry season migrants from the rural areas of northern Nigeria do not have a Western education. Consequently, most factory workers from the southern parts of the country tend to have more clerical and technical skills than their northern counterparts (Lubeck, 1986). In recent times, however, the educational gap between the north and south is narrowing and there are increasingly large numbers of Hausa and northern employees with clerical and technical skills in the factories. This conjecture is further supported by the evidence presented in the three case studies.

Mobile Labour and the Business Sector

It is necessary to reiterate here the importance of an adequate and steady supply of labour to the growth and development of a business environment and for the firms operating in such an environment. The survival of a business enterprise is sometimes threatened by a highly mobile labour force or labour that is in perpetual flux. Few studies have examined the implications of labour migration on businesses in Nigeria's organised private sector. In some of these studies, the implications of migration have been restricted to rural agriculture (Olusanya, 1969; Osuntogun, 1976; Makinwa 1981; Essang and Mabawonku, 1974).[196] As such, there is a veritable gap of information on the effects of migration on the Nigerian business sector. The sequence of events that engender migration and the range of characteristics involved create methodological difficulties in evaluating the effects of migration on business development and rural stagnation.[197] However, it may suffice to mention that the cities and urban areas that act as hosts or recipients of migrants experience severe pressure on infrastructural facilities.

The influx of migrants has led to the development of periurban settlements around the city of Kano in areas such as Dakata, Brigade, Gwagwarwa, and Kundila. In addition, this has led to a boost in the mortgage business and spiralling rent increases in the housing sector. In their response to the windfall in the housing sector, many indigenous merchants of Kano, including an indigenous Hausa proprietor of a plastic factory covered by this study, diversified a considerable amount of their wealth to the housing sector. The trend has resulted in several hurriedly and poorly constructed houses for the jobless and low-income earners in areas such as Brigade, Gwagwarwa, Tudun Wada, and Dorayi.

Most factory workers in Kano are migrants. A profile of the factory employees in three plastic firms in Kano revealed that 114 of them, representing approximately 69 percent, were migrants from other parts of the country. Significantly, only 31 percent of the male employees in the three factories were indigenous to Kano. The plentiful supply of migrant labour continuously creates surplus and cheap labour for the business enterprises operating in Kano and other parts of Nigeria. Piore's (1979) assertion that temporary migrants often accept dead-end jobs that indigenous people would reject has some validity regarding the evidence presented in Kano. Thus, for many temporary migrants from rural areas, the prime desire is to find capital from whatever source and then invest it in petty trading or on the farm.

There are significant links among migration and issues such as voluntary labour turnover, absenteeism, and lateness to work that cannot be easily glossed over. Human movement involves costs and benefits for individuals, groups, and the economy of a region (Amin, 1995, pp. 29–40).[198] Studies of the effects and implications of labour migration in Nigeria (Osuntogun, 1976; Makinwa, 1981; and Olusanya, 1969) agree that out-migration of energetic youth from rural areas adversely affects agricultural production, which usually results in retarded development and rural stagnation. The erratic movement of migrants during certain times of the year makes the subject of migration relevant to the discourse of business activities in Kano.

In the factories, most of the casual employees are migrants. Because of the highly mobile nature of migrant labour, especially from northern parts, business activities in Kano experience the toll of erratic labour movements and shortages during certain times of the year. During the rainy season, the casual or surrogate labour in the factories from rural areas usually deserts the cities to cultivate arable fields abandoned in the dry season. Casual or surrogate employees often develop the feeling that they do not owe employers any obligations and can therefore withdraw their services at any given moment.

The manner in which casual migrant workers desert firms is particularly worthy of mention. Since most casual hands in firms are uneducated, their movement is often without formal written notice to their employers. This inevitably leaves most operators of firms unprepared. Entrepreneurial preparation for unforeseen movements of employees is made difficult because of the erratic nature of migration. The semi-arid conditions of Kano and its neighbouring districts make rainfall a highly cherished commodity. A heavy downpour of rain may immediately trigger a mass movement of casual labour to the countryside, leaving most businesses paralysed. In Lebanese firms, where most workers are hired on a daily basis, the effects of mass departure of migrant workers to rural areas during the raining season are strongly felt. This behaviour is typical of agriculturally oriented employees from the northern Nigeria. Employees from the southern parts of the country tend to rely more on wage employment in the modern sector.

The evidence from the three plastic factories also reveals that more than half the employees in the Hausa factory employees with rural origins commuted weekly between the city and their rural homes. Approximately 68 percent of the factory labour-force that were not indigenous to Kano visited home at least once a year. The reason for this was to join parents, spouses, and children left behind in the villages. The weekly movements by the local workers, though largely unnoticed, exercised significant side effects on business operations. Islamic norms make the keeping of women in seclusion, or *kulle*, a strong practice in the predominantly Islamised parts of northern Nigeria. Some highly religious low-income factory workers from rural areas see the city with its attractions as corrupting to the womenfolk. Consequently, wives are left behind in the villages. Employee movements during the weekends to the rural areas become the established link for rejoining the family. One of the implications of the weekly movement of labour from the organised private sector, is the scarcity of labour in factories during the weekends. Sometimes there are labour shortages in factories where overtime work may be needed to make up shortfalls in production levels.

The strong attachment of most Nigerian workers to their homes comes from social links such as marriages, and the need to make remittances to idle relatives at home. The migrant worker constantly struggles to overcome a 'financial burden' which his 'privilege position' as a salaried worker or 'wage earner places on him'. This dependency burden, which is not restricted to immediate family members, includes those left behind in the villages who may either visit the worker in the city or request that help be sent to them (Yesufu, 1962, p. 57). Although only 51 percent of the factory employees interviewed were married, nearly all of them supported at least one relation at home. The results of our study are supportive of Yesufu's (1967) study concerning migrant remittances to rural homes. The flow of wealth to the older generation in the rural areas usually comes from the much younger migrants in the cities (Caldwell, 1976).

In many African societies, including Kano, the most successful migrants are those who are constantly in touch with home. For Yesufu (1962), employers have to accommodate migrant worker's love for their village homes. Societal values and norms are important in reinforcing the migrant's links with his traditional village roots. Gugler (1965) observes that the attachment between most urban workers and the village exists because 'they care about the opinions people in the villages hold of them'. Expatriate businesses sometimes offer sanctuary for employment. In Kano and parts of northern Nigeria, expatiate businesses are found in both the formal and informal sectors of the economy.

Expatriates Investors and the Socio-Cultural Milieu

The expatriate community in Kano and elsewhere in Nigeria is relatively small. Although the exact population is not known, there are probably over 80,000 expatriates and their families residing in Kano. The economic activities and incomes of most expatriate groups including, Europeans, Arabs, Lebanese, Syrians, and Asians put them among the urban upper class. Lebanese and Syrians business community in Kano overlap within the category of expatriates and indigenous people. Most of the Lebanese and Syrians are second-generation citizens, which automatically allows them to take up Nigerian citizenship. Others, and especially newcomers, often retain their Lebanese, Syrian and Indian citizenship. The Kano Club and the Bagauda Holiday resort often provide ideal places of leisure and forum of interaction between expatriates and their local business counterparts.

There is no single, but several ways by which investors' react to the socio-cultural milieu. Initial challenge posed by competition with local business groups as noted elsewhere in this study encourages social networks and group formation among expatriate investors. However, some expatriate groups are more assimilable than others are. The Lebanese and Syrians for instance are more assimilable than European and Asian business groups.

In spite of the low participation of Muslim women in the open urban market, many Lebanese and Syrians textile traders in Kano conduct their daily activities with their wives and children. Like their Lebanese and Syrian counterparts, Chinese operators of hotels and restaurant business in Kano often employ the services of family members. Nevertheless, many Lebanese factories, including the Challawa plastic factory covered by this study, do not have female workers on the shop floor.

Although the local difficulties confronting both indigenous Nigerian and expatriate businesses are similar and appear to be enormous, expatriate businesses appear to have the upper hand in a number of key decisive areas. First, expatriates' access to information in foreign markets puts them ahead of their local counterparts in economic ventures. Such access include reliable supply lines for high-quality and cheaper industrial raw materials and spare parts. Expatriate businesses tend to possess the relevant material and technical skills necessary for the operation of their firms, whereas in Kano and other parts of Nigeria, a large chunk of indigenous technical skills, as one expatriate manager observed, 'are better suited for building and road construction rather than the manufacturing industry'.

Among many expatriate business operators, the circumstances that often trigger immigration include a strong urge for success against all odds. Lebanese investors have particularly proven that immigrants possess patterns of highly adaptive economic behaviour. The development of positive economic attitudes in foreign lands often makes expatriate businesses more successful than local ones. Among the Lebanese and Asian communities, the high level of emotional support and informal capital raising networks for helping new settlers are among the factors that have contributed to expatriate economic progress in Kano.

Expatriate businesses in Kano and elsewhere in Nigeria sometimes do economically better than their indigenous counterparts because of the different social conditions under which they operate. Unlike their local counterparts, expatriate entrepreneurs and managers are often spared the ordeal of primordial financial pressure. This implies that some unnecessary social expenses incurred by indigenous entrepreneurs and managers are often absent or limited in expatriate firms.

Managerial self-assessments in the three plastic factories were at best parallel to what many employees said about them. The personal assessments were also in contrast to our research findings. To a large degree, employee assessment of both indigenous and non-indigenous expatriates managers were based on the peculiar circumstances under which they found themselves. Table 8.2 represents employee-matched assessment of indigenous and expatriate business operators. This comparison is based on employee response to an open-ended question: 'what do you think are the similarities or differences between

Table 8.2. A Matched Employee Assessment of Business Proprietors

Hausa Managers	Lebanese Managers	Chinese Managers
1. Less collectivist	Fairly collectivist in business	Fairly collectivist in business
2. Undisciplined	More undisciplined	Highly disciplined
3. Suspicious of workers	More suspicious of workers	Most suspicious of workers
4. Empathic	Fairly empathic	Less empathic
5. Fairly open and friendly	Less open but friendly	Less open and unfriendly
6. Tolerant of mistakes	Fairly tolerant of mistakes	Less tolerant of mistakes
7. High job insecurity	Prone to a higher rate of job insecurity	Prone to the highest rate of job insecurity
8. Lax control	Stricter control	Strictest in controls

Source: Author's field interview 1995.

indigenous Hausa, Lebanese and Chinese business operators'? The factory workers perceived Chinese entrepreneurs and managers as highly disciplined but less empathic, and more suspicious of subordinates. Although empathic and tolerant of mistakes, indigenous Hausa and Lebanese business operators were regarded by their employees as lacking in self-discipline. Factory workers also saw Lebanese and Chinese business groups as more collectivist[199] than their indigenous counterparts. The attribute of collectivism is a common characteristic of immigrant business groups.[200] On a general note, the employees in the three factories saw Hausa, Lebanese, and Chinese firms as prone to higher rates of job insecurity. Nevertheless, Chinese business operators were seen as worst offenders as far as job insecurity was concerned.

Unlike indigenous firms, expatriate and multinational firms place more premiums in recruiting employees with skills and qualifications than on social background. Evidence in the Chinese plastic factory has shown that most of its workers were not indigenous to Kano. Although Nigerian labour provisions require that preferential treatment be given to people indigenous to the place where a factory is located, expatriate firms seldom meet such requirements.

Despite the absence of employee training programmes, expatriate and especially multinational firms perform better in the area of employee training than their local counterparts. The absence of training programmes in indigenous and Levantine firms sometimes accounts for employee incompetence and misuse of resources, which often culminate in their poor performance. Invariably, the growth of the modern manufacturing firm thrives on skill and competence of its employees. Observation of indigenous manufacturing firms in Kano has shown that the management cadre of most Hausa factories was composed largely of kin members who were often retired civil servants with no experience in manufacturing activity. This scenario led to frequent hitches in the maintenance of regular supply lines for industrial raw materials both locally and from overseas countries.

Much of the evidence in the northern Nigerian plastic industry shows a low level of trust among employers of labour and factory workers. It would be recalled that the subject of trust was discussed in Chapter Four. However, the level of trust also observed, has an inherent connection with entrepreneurial attitudes towards the future and time in general. This leads us to the discourse about time. Luhmann (1979, p. 10) observes that time has a problematic relationship with trust. He maintains that to show trust is to anticipate the future or behave as if though it was certain. Conversely, a pervasive sense of distrust has less regard for time and particularly the future.

Cultural Conceptions about Time

Research about entrepreneurial response to 'business time' has recently gained currency among social scientists and management scholars as a medium through which cross-cultural business attitudes can be discerned. Time is the medium through which cultural events take place (Archer, 1989). Time is also relevant because it possesses the potential for shaping economic and social habits in any given society. For Giddens (1995, p. 17), there is an intimate connection between time and modernity. A study on entrepreneurial attitudes in a developing society via perceptual processes therefore requires a discourse on time as a cultural variable.

The different conceptions of times in academic discourses include, cultural, religious, social, and economic. In this study, the cultural, religious, and economic dimensions of time are examined. Due to the complex nature about cross-cultural reflections on time, effort is made at reducing the abstract notion to a simple understanding. In business analysis, the cost element, often associated with time, implies that the optimum use of 'economic time' is very important to business performance. Efficient utilisation of time is often associated with economic success. In manufacturing activities where quantity, quality, and operational targets are important yardsticks for measuring success, efficient use of time and expedient execution of tasks combine to enhance the attainment of goals and maximisation of productivity. In contrast, lax attitudes toward the use of time often yields low productivity, losses, and the disappearance of manufacturing firms in most developing countries.

Manufacturing and other forms of processing activities require precise scheduling to ensure that diverse activities are smoothly channelled through the production process. Thus, for Inkeles and Smith (1974, p. 160), the necessity of good time management is 'most evident in the assembly line, since it rigorously imposes the necessity that everyone starts and stops at the same time, that each process be allocated a precise amount of time, and that each step be completed as scheduled'. Effective use of the time in business connotes reliability or predictability. The ability to fulfil customer needs at the appropriate time goes a long way in the reinforcing mutual trust among business partners. Lack of entrepreneurial discipline and the failure to follow time agreements in the supply of goods and services can cause a loss of credibility and lower patronage by customers. Thus, entrepreneurial attitudes, planning, and investment schedules are intricately linked to their perceptions and attitudes in relation to time, past, present, and future.

Societies have various ways and standards by which they reckon and make uses of the time at their disposal. There are also cultural differences in tolerance for punctuality and lateness. While issues about punctuality are taken seriously in the capitalist world economy, in less industrialised societies, they are often trivialised or glossed over. It was therefore common to hear employees in the plastic factories make a distinction between 'Nigerian', 'African', and 'European times'. The latter is conceived as mechanistic, mathematical, and punctuality-conscious, while the former was perceived by factory workers as relaxed, non-mechanistic, and lacking punctuality-consciousness. Consequently, while running thirty minutes or an hour behind schedule may not matter in the 'Nigerian time' context, it does matter in the capitalist socioeconomic sense.

There are, therefore, differences for the variations in time which may be acceptable (Adler, 1991). A 1976 UNESCO report on societal use of time shows that some societies or cultures are 'more pathologically time conscious than others'. Although indigenous Hausa business people are very alert to market opportunities and economic windfalls associated with short-term shortages, an observer of their activities would notice their lack of regard for 'economic time' or precision associated with the modern capitalist mode of production. This lax attitude towards time is not peculiar to Kano but is a common pattern among business groups in developing societies.

The various Hausa conceptions and qualifications of time are derived from three sources: Islamic, indigenous values, and elements of Western thought based on the Gregorian Calendar. This implies that, in Hausa societies, Arabic, Hausa, and English terms associated with time are used interchangeably.[201] There are elements of pluralism in Hausa time reckoning, with varying degrees of emphasis on cultural and religious notions. Generally, Hausa entrepreneurs operate with two calendars, the Islamic calendar, based on the lunar month, and the Gregorian calendar. The two do not often match. The use of multiple and incongruous calendars and the attempt to synchronise them often gives the impression of two different years in one. Cross-cultural variations about time imply that Hausa definitions of time are neither the same thing as the English 'time', nor the Swedish '*tid*'. In business dealing with foreigners, multiple references to time creates ambiguities that frequently lead to cross-cultural mistiming. Generally, the African idea of time is phenomenal rather than mechanistic. This implies that, 'time is noted in terms of the individual's phenomenal experiences of concrete events, rather than in terms of mechanistic processes characteristics of clocks'.[202] In many African societies, including Kano, time is linked to the daily movement and positioning of the sun and the seasonal variations of wet and dry seasons.

Hausa linguistic qualifications about time include today, or *yau,* and *locaci*, which implies a particular time. The term *sa'a* implies an hour or time. The lunar month, or *wata,* is a significant point of reference for business people in their daily business transactions. In recent times, the *watar-Bature*, or European month, has become the basis for calculating wages and salaries in both the informal and formal economic sectors. Other Arabic notions and variants of time used in Hausa society include *waqt* and *zaman* or *zamani*. The term *waqt* in Arabic implies 'a specific point in time',[203] whereas *zaman* which is general, denotes an unlimited, or macro, time frames. Consequently, the term *zaman* has a historical reference, and it is commonly used in astrology. In Hausa terminology, terms such as, *rana*, and *locaci*, which have short-term meanings, are within *zaman*, a rather long-term notion of time. Thus, 'instant time' or 'time-now' usually signifies prayer time. In the instant time, man is left 'without past or future' (Gardet, 1976). Societal attitudes regarding punctuality and lateness are based on cross-cultural interpretation and understanding of these concepts.

There are two major schools of thought concerning Islamic views about time. The first school, which postulates a circular notion of time, comes from the works of 'Al-Kindi (an Arab philosopher), Farabi (a Turk), Ibn Sina (a Persian), Ibn Tufayl and Ibn Rushd (from the Maghreb)', respectively.[204] Their circular views of time were presumably derived from philosophy or *falsafa*. In their schema, the movement of time is a continuous process oscillating back and forth through the seasons. The second school, which is a variant of the circular notion, perceives time as linear. The Arab scholar Imam Ghazali (1058–1111) observes that the theory of eternity of time in the world was sacrilegious or impious and contradictory to the teachings of Islam. It is worth mentioning that the two parallel views championed by these Muslim scholars do not foreclose the Islamic notions about spiritual and practical times.

The spiritual, or unknown, time connotes secrecy, or *gaib*, the knowledge of which rests only with Allah the Supreme Being. Given this fundamental view, any reference to a possible future is qualified by the formula 'if God so wills' or *insha Allah* (Gardet, 1976, p. 204). Knowledge of the hidden future, or *gaib*, is associated with periods dealing with life or death and the 'final hour', or day of judgement. The practical time is one in which human activities are submerged. Time is mysterious, which implies that, despite its universality, it defies adequate explanations. Thus, a day, in spiritual time, may be equivalent to several thousand years 'on the plane of this life'.[205] The practical or non-spiritual aspect is useful for our analysis. The spiritual reference to time, which overshadows its daily practical usage, in Hausa society often gives the erroneous feeling that, because 'time is hidden', it is not worth planning for. The consequence of which may be a lack of regard for punctuality and reliability in most Hausa communities.

Paradoxically, Islamic and Hausa cultural values do not seem to encourage the abuse of time. It is particularly held that 'nothing is more destructive to the human being than the abuse of time' (Al-Bukhari, Vol. 6. 1987, pp. 267–351). Hausa conception of time, or *lokaci*, tacitly implies 'promise', or *alkawari*. Consequently, time that is linked with promise, or *alkawari,* is synonymous with a load, or *kaya,* which must be disposed of at an appropriate place. Reneging on a promise, or *warware alkawari,* which is tied to the notion of time, is seen as a sign of personal weakness. Poor attitudes regarding time may seem to be a reflection of the erosion of cultural values. Perhaps the poor attitudes towards time in developing societies are due to the absence of the development of indigenous technologies associated with chronological time.

In Islamic societies, the day is divided into five periods corresponding to prayer times. Entrepreneurial appointments and business schedules in Kano are usually linked to meeting before or after prayer times. In Greek texts, the day and night were divided into three periods (Leoyd, 1976, p. 119). Islamic knowledge about time is practical and based on the different positions of the sun, and moon at a particular time of the month. The various positions of the sun and the elongation and shortening of human shadows during the day determines specific prayer periods and the time for breaking the fast, as happens during the month of *Ramadan.* This implies 'respect for collective rhythm' which is neither 'excessive haste nor sluggishness', but is simply a question of being in the proper place at the proper time (Bourdieu, 1977, p. 162).

Entrepreneurial Use of Time

Societal response to time in many African countries is that of tolerance and flexibility. This scenario contrasts with the capitalist notion of time, which is mechanistic, mathematical, and rigid. A day in Hausa is called *rana*, which literally translates as 'sun'. This tacitly implies that a day in Hausa culture begins at sunrise, whereas in Christendom and in other parts of the capitalist world, the day starts at midnight. Thus, across cultures, terms such as day and night have different connotations due to variations in their starting and ending points. Although entrepreneurs across different cultures may use the same calendars or wristwatches, their references to time often differ, even in the same environment. Callaway (1987) observes that, in Kano, most foreigners, especially Europeans, are generally considered 'always in hurry', due to their respect for chronological time. Among indigenous entrepreneurs in the private business sector, there is no scheduled time for beginning or closing business activity. Their diversification of resources in the running of different unrelated business activities often involves tremendous dissipation of energies and considerable amounts of time.

The Hausa cultural emphasis on altruism cause indigenous entrepreneurs and local employees in modern factories to constantly struggle to keep a balance between the use of social and economic time. Social or potential time, in this context, refers to that which is inevitable, which 'falls within the rhythm of natural phenomena' and is linked to communal expectations.[206] Timely arrival at work depends on events in the neighbourhood. A social event in the local community, such as a funeral, marriage, or child-naming ceremony might delay or prevent the entrepreneur or an employee from going to work. These attitudes are reinforced by strong notions of collective responsibility to the society. Moreover, entrepreneurial attitudes towards time are embedded in cultural receptiveness about lateness. Thus, indigenous entrepreneurs and their managers have been generally observed to be more receptive and tolerant of workers lateness than their expatriate counterparts in both commercial and manufacturing activities.

Lubeck's (1986) study of firms in Kano reveals that, in some expatriate firms, managerial insistence that indigenous Muslim workers complied with the capitalist notion of time generated much tension and was strongly resisted by the workers. A problem relating to indigenous factory employee confrontations with the capitalist conception of time was encountered at the Chinese factory. In the Chinese plastic factory, the use of punch cards was viewed by workers as very harsh and was strongly resisted, with several attempts by the employees to destroy the time machine. In spite of this, entrepreneurs and workers in the industrial sector feel the pressure of time more on day-to-day, and even hourly basis because of the requirement that the activities of many others are coordinated simultaneously to ensure smooth operations (Inkeles and Smith, 1974, p. 162).

Cultural conceptions about past, present, and future exercise strong influences on entrepreneurial investment policies. The investment patterns in future-oriented cultures are based on long-term economic projects, whereas in present-oriented cultures, business emphases are placed on speculative short-term economic ventures. In most African cultures, including the Hausa, the conceptions about time are a combination of past and present with less emphasis on the future.[207] There are elements of traditional and modern insights in Hausa time perception.[208] The smaller emphasis on the future has made late scheduling and poor forecasting to become integral elements in the operation of indigenous factories.

To invest in the future is to exhibit trust in a particular system. Conversely, entrepreneurial lack of commitment for the future implies a lack of trust. Indigenous entrepreneurs often do not invest in the future because of the low trust they have for the economic system in which they operate. Adler (1991) observes that future oriented cultures like Japan invest a lot in staff training and the development of employees. This is reflected in the Japanese work ethic and the concept of lifetime employment. Managerial attitudes towards time in the Lebanese plastic firm were typically present-oriented. This led to reckless planning with no training programmes designed for employees. In Kano, as in most parts of Nigeria, the absence of lifetime or stable employment in business enterprises does not oblige employers of labour to invest in employee training.

A study of Chinese culture by Mathews and Mathews (1983) reveals that, unlike the Japanese, the Chinese are past-oriented. Consequently, Chinese attitudes towards modernity and innovation depend on how they fit with their past experiences. Their findings contrast with the results of our study, although the impact of the foreign environment cannot be entirely divorced from the Chinese management strategy in the Kano plastic firm. The long working hours showed that the Chinese who were more present-oriented maximised the time at their disposal without recourse to its implications for employee fatigue or boredom. Furthermore, the absence of formal training programmes in the Chinese factory also showed a lack of planning for the future and a disinterest in employee retention.

Thus, while indigenous Hausa entrepreneurs were strongly past-oriented, their Chinese and Lebanese counterparts were predominantly present-oriented. Generally, entrepreneurial and managerial attitudes regarding time in the three plastic factories were poor and hardly in consonance with the demands of the modern manufacturing sector.

Summary of Part Two

I started the Second Part of this study by looking at the research setting. A case by case analysis of the three plastic firms was presented with a summary of findings in each firm. This was done by first looking at the economic environment in which the firms operated. Through simple cross-cultural comparisons of three case studies in Chapters Five, many similarities and differences in managerial styles have been illuminated. Recruitment strategies through social networks in all the firms were seen as favourably disposed towards males, siblings, and kin groups, while they discriminated against female employment. It was observed that informal recruitment of employees through social networks inhibited bargaining procedures and constrained the proper functioning of employee associations in the plastic sector.

In the Sixth Chapter, the entrepreneurial profit motive was examined. Entrepreneurial cultural definitions, calculations, and the use of profit were discussed extensively. It was observed that reckless management or use of profit constrained business success. Furthermore, employee views regarding profit sharing in the firms were often spurious and in conflict with those of management. Employee wages in the plastic firms as revealed by this study, were however, very low and hardly at par or indexed against national inflation. The payment of low wages and salaries as shown by the study, engendered discontent among the rank and file of workers in nearly all the plastic firms.

The focus of analysis in Chapter Seven was on entrepreneurial and managerial control methods. Entrepreneurial control in the indigenous Hausa firm was patrilineal and less bureaucratic than in the Chinese and Lebanese firms. The administrative permissiveness of the indigenous entrepreneur resulted in a smooth working environment. In contrast, the Chinese firm was very autocratic. Stringent controls in the Chinese firm were responsible for tensions among employees and supervisors in the shop floor. Compared to Hausa and Lebanese managers, the Chinese were more task-oriented as shown by the twelve daily working hours in the Chinese plastic factory. Lebanese managers, on the other hand, were less autocratic than the Chinese, but stricter than their Hausa counterparts in the control of shop-floor employees. Managerial treatment of longtime employees as casual labour, and payment of arbitrary wages among workers performing similar tasks, were some reasons for employee discontent in the Lebanese factory.

To a large degree, poor educational qualifications of most workers in the plastic industry were explained as a major reason for their abuse by their employers. In view of the absence of knowledge on bargaining procedures among employees, wage determination in the factories was unilateral and different from the bipartite or tripartite requirements that exist in the Nigerian industrial legal system. The recognition of employee unions by Chinese management does not imply that they were more tolerant or democratic than their Hausa and Lebanese counterparts. Observation of the Chinese factory reveals that the worker's union was a camouflage aimed at giving outsiders a picture of seeming fairness. The most common similarities across the three plastic firms were the absence of training programmes, poor working conditions and incredibly low wages and salaries.

In Chapter Eight, informal sector economic activities were examined in greater detail. Urban surplus was seen as a product of institutionalised migration. The Chapter analysed the patterns of seasonal labour fluctuation, in particular, and their impact on the business sector in Kano and northern Nigeria as a whole. It also attempted to show *inter-alia* the link between the use of migrant labour and the problems of high attrition experienced by many Nigerian firms during certain times of the year. In particular, the informal sector of the economy, which is characterised by easy entry, is used as a training ground by migrants seeking employment in the organised industrial sector.

The modalities by which the expatriate community in northern Nigeria react to the socio-cultural milieu was discussed. Business groups of Middle East origin especially Lebanese and Syrian were seen as more assimilable than other expatriate groups. This characteristic of easy assimilation into the local community was partly attributed to the Islamic religion to which most of the Lebanese belonged. Entrepreneurial attitudes towards employees showed they were generally less future oriented. This led to the discussion about time and its relation with trust. Entrepreneurial low trust in the future was largely reflected by the absence of external training programmes for factory employees.

Part III
General Summary, and Conclusions

9 Women in a Hausa Society

> Work might be performed for the state, a capitalist employer, a feudal lord, a husband, or collectivity. It may be paid or unpaid, create use or exchange values. And it might be undertaken on a slave plantation, a commune, or in the 'public' economy of the market or the 'private' economy of the household (Glucksmann, 1995, p. 69).

This chapter examines the role of Hausa women in the northern Nigerian economy. The focus of analyses is on the historical, cultural, and economic dimensions of women's contributions to society. In the first section of the Chapter, female seclusion is examined from a northern Nigeria perspective. The second section deals with female labour participation in the formal industrial sector. The final section of the Chapter compares female household economic activities with industrial labour. An analysis of female economic activities is crucial in providing additional clarification on the subject of economic rationality discussed in the second Chapter of this study. In particular, the relative absence of Hausa women in the formal or open economy provides deep-seated understanding regarding the economic rationality of Muslim Hausa society.

Female-related discourses are at best inconclusive. Universal theories of female subordination, class and biological analyses have become increasingly controversial. This polarisation of research foci has further compounded the problem for investigation in the area. Nevertheless, much of the literature on Muslim women in purdah and veil paints a pervasive picture of oppression, subjugation, and retardation. Muslim women are frequently perceived as objects of pity often holding marginal and counterproductive statuses in their societies. For Coles and Mack (1991) the role of Muslim Hausa women reinforces the idea of gender asymmetry. Hill's (1977, p. 84) description of the thrusting of 'little pre-nubile girls' into full purdah without access or contact with the outside society, presents an awkward portrait of female subjugation and misery in Kano. Many Western scholars see the practice of purdah in northern Nigeria as a restriction of female liberty.[209]

The preponderance of evidence of Islamic teachings and Hausa cultural values show the claims about female subjugation are very often unsupportable. Today many Muslim Hausa women wear veils and maintain purdah rules out of conviction rather than compulsion. Hausa women balance both Islamic and cultural expectations with choice, and are therefore not pawns in patriarchal Muslim society as often seen by many outsiders. For the Muslim Hausa women, keeping purdah rules and wearing veils are symbols of religious identity and signs of protection, and respect rather than oppression.

Today, female seclusion and wearing of the veil in Hausa society have become proofs of religious piety. Maintenance of purdah rules and wearing of veils is often the distinguishing symbol between the Muslim and non-Muslim women in Kano and parts of northern Nigeria. The veil is therefore functional in creating a religious identity and joining the Islamic community into a cohesive unit (Powell, 1982; and Almunajjed, 1997). In many Islamic societies, the veil is considered a shield from the 'adulterous eye'.[210] Among Muslim women, the veil is an external precautionary safeguard that symbolises virtues of conduct. The Qur-an enjoins compliance with the norms of religious modesty. Muslim females are admonished to cover their bodies completely with the exception of the face and the palms.[211] Religious garments are therefore instruments of modesty, which also hide the body (Boudhiba, 1982, pp. 49–50). A descent dress is an essential part of piety and good manners in Islam.[212] Barkow (1972), for instance, perceives the practice of purdah in northern Nigeria as 'permitting women to acquire individual wealth while increasing the male burden'.

That women constitute nearly half the population of Hausa society is hardly controversial. It may therefore seem to some observers that their low participation in the formal or open market economy may lead to stagnation and decay of society. Nevertheless, the evidence in many parts of northern Nigeria, including Kano, presents a contrasting picture. Misconceptions about the role of women in the Islamic society can only be extirpated by differentiating between the teachings of Islam, and local customs and traditions which are often erroneously conceived as part of it (AlMunajjed, 1997, pp. 31–32). From the Islamic point of view, women are considered equal to men in legal, political, economic, and social life. This implies that despite the obvious biological differences both men and women are spiritually equal in Islam.

The literature on gender roles in Hausaland before the advent of Islam is rare. Empirical analysis about the role of women is either weak or scanty. Probably, more than half the history of Hausaland might have been lost or was never chronicled before the encounter with the outside world. Most of the accounts on the role of women before the adoption and spread of Islam are oral or evidence from archaeological materials. Consequently, an analysis about the women's activities in Hausaland before Islam is a complex exercise. Such an exercise is however, worth pursuing because it provides clues that demystify misconceptions about the role of Muslim women and Hausa business rationality.

The most vivid accounts of illustrious Hausa women are documented in the *Kano Chronicle* and the *Infaq al-maisuri* of Sultan Bello. From these sources, it is now known that many great women in Hausaland such as the queens of Daura-Daurama, and queen Amina of Zazzau held positions of leadership and played significant roles in patriarchal societies for many years.[213] Historical accounts also provide evidence about the role of women in the Borno Empire and Nupe kingdoms. From historical accounts, folktale and traditional songs, it is known that Hausa women were not secluded, before the adoption of Islam. This makes it incumbent to examine the historical evolution of female seclusion in northern Nigeria.

Female Seclusion, a Historical Perspective

The social, economic, and political roles of women have passed through changes under three different historical epochs. First is the role of women in the pre-Islamic or *jahiliyya* period. The pre-Islamic era refers to early Hausa History up to A.D.1500 (Paden, 1973, p. 45). Second is the position of women during the early years of Islamisation, and thirdly, women after the reformist jihad. The pre-Islamic era marks the period in Hausa history when female seclusion was absent. The *jahiliyya* was characterised by permissiveness and mixing of all sexes in the daily social life.

During the early years of Islamisation, relaxed forms of seclusion practices were introduced in the local communities. Seclusion, or *kulle,* during this period was weak and largely confined to a few Muslim clerics or *malams* and the ruling classes. Male or spousal supervision of female compliance to seclusion rules was weak and sometimes absent. Part of the reason for the Islamic jihad in Hausaland was to eradicate syncretic practices including the free mixing of sexes in the social milieu. Some of the early studies of Hausaland by Western scholars describe the various economic activities of women in the open market, which were challenged by Islamic *jihadists*.[214] Strict enforcement of seclusion practice came during the Muslim-Fulani led jihad in 1807. The *jihad* marked profound changes in the social, political and cultural conditions of Muslim Hausa women (Hiskett, 1973; Smith, 1981; Boyd and Last, 1985).

The practice of *purdah* or keeping women in seclusion in northern Nigeria is synonymous with the introduction of Islam. Today Hausaland is one main area in the whole of Africa where the strict seclusion of adult females is a religious norm. There are however, variations on the extent of strictness, between rural and urban areas. In the rural areas the homogeneous nature of the society and the demands imposed by the agricultural sector often makes the practice weak. Most rural women seek jobs as paid farm labourers in harvesting or threshing harvested grains. Coles and Mack (1991, p. 8), observe in their study of rural Hausa women, that only religious teachers, wealthy farmers and merchants can afford to and do seclude their wives. In the cities however, the relative affluence of urban dwellers, economic independence, and heterogeneous mixture of the society encourage seclusion. Among educated women in the urban centres, seclusion rules are not always adhered to because of the obligations placed on women by job requirements in the formal sector.

This significant socio-cultural transformation in the lives of Muslim women in northern Nigeria did not come without some initial resistance among the indigenous people. The enforcement of women restriction often resulted in friction between Muslim Fulani evangelists and Hausa society (Hiskett, 1973). The practice of purdah was initially resisted in some communities due to a misunderstanding that the practice was a Fulani imposition. Acceptance of seclusion rules by Hausa women was initially complicated by the fact that pastoral rural Fulani women who natured livestock and hawked milk did not practice purdah or *kulle*.

A look at the lives of Muslim women in urban Kano shows that seclusion has not exacerbated their plight, rather it has provided them with a descent and secured position. The Islamic injunction that 'Paradise lies at the feet of mothers'[215] underscores the importance of women to society. Islam encourages the extension of more kindness and support towards mothers than fathers. Islamic prohibition of free mixing of the opposite sex aims to create a pure society while preventing any evil or negative consequences which may arise from such mixture. In Islam, the family is the bastion of Society. A shattered family therefore implies a troubled society. Regulation of free interaction between sexes in Islam is considered a basic step to enhancing family fidelity. Seclusion introduced structural changes in northern Nigeria. Writing about seclusion in Hausaland and the fundamental changes is brought to the productive roles of women, Barkow (1972), posits that:

> Islamisation provided Hausa women with an opportunity to forsake labouring in the fields and to develop craft skills. Men encouraged or at least accepted wife seclusion because it increased individual prestige and was associated with piety. This increased the men's burden and permitted women to acquire individual wealth. Since much of their activity involves the preparation of cooked foods, an economy based on high degree of labour specialisation arose (Barkow, 1972, p. 327).

In discussing gender issues in Hausaland, it will be necessary to exercise caution on over generalisations. This is in view of the significant differences between urban and rural women. Even in urban centres, the differences in the socioeconomic life patterns of aristocratic, merchant, educated and uneducated urban proletariat in the commercial and industrial sectors and working class women is great. The observance of seclusion rules coincides with class and educational distinction. Educated urban women in Kano today can be found in public positions such as commissioners, leading newspaper editors, medical doctors, nurses, managers of banks, and factories.

Many urban women live in conditions under which seclusion is difficult to maintain, they share compounds with unrelated families and find mobility necessary to carry out income earning subsistence (Schildcrout, 1983). Sharing compounds with unrelated family members is often the reason migrant Muslim and Hausa women in parts of cities like Sabo-Ibadan, Agege-Lagos, or in the *zangos* of Accra, and Kumasi, find it difficult to keep seclusion rules. Conversely, migrant Hausa women like their female counterparts are very active in open market retail trade in these areas. In parts of Nigeria and the African continent, market women control retailing activity (Simms, 1981, pp. 141–168). Usually, acquisition of outside norms during contact with non-Islamic cultures is another factor that paves the way to non-adherence to seclusion rules.

In recent years, female seclusion has become a standard assessment for male economic success and independence. Seclusion implies that a man is sufficiently wealthy enough and able to meet his family and wife's needs without her going out to work (ibid.). Conversely, free movement, not only violates religious norms; it can also reflect male economic failure. Nevertheless, divorced women, or *bazawarai,* may live in their family compounds and engage in economic activities. Although educated Hausa women may find employment in public places; few of them work in the industrial sector.

Among the factors which account for the low participation of Hausa women in the industrial sector include, cultural, religious and logistic factors. The indigenisation policies of the state governments in the early 1970s did not make adequate logistic provisions consonant with Islamic practice to facilitate female participation in industrial labour. Part of this is the absence of separate facilities and congenial working environments for adult of married Muslim women. The lack of separate facilities in work places in northern Nigeria will most likely keep female participation in the formal economy low.

Female Industrial Labour

A common characteristic of business firms in patriarchal societies is the overwhelming disposition towards male employment. The definition of gender roles becomes more rigid in masculine rather than in feminine societies (Hofstede, 1980). In masculine or patriarchal societies, the lines about gender roles tend to be sharper and more clearly defined. Gender-related issues have implications for employee industrial relations in firms dealing with opposite sexes (ibid.). Some highly masculine societies in the world today include Austria, and Japan, whereas Sweden is generally regarded as a feminist society.

The Hausa society of Kano is patriarchal with strong patrilineal tendencies, in which men who are the guardians of the family, provide both economic and social security to their womenfolk. Practical Islamic taxonomies and Hausa cultural ethics require that men, and not women, should work to feed or provide sustenance for the family. The role of Hausa women in the formal industrial sector is quite small. This, however, does not imply that secluded women who operate outside the formal economy do contribute to the general economy. To a large extent, they do.

Socialisation into gender roles among the Hausa often begins at the tender age of three to four years. The societal understanding of the role of women in the economy derives from cultural and religious teachings.[216] Sometimes, adequately protected women do not seek employment in the 'open economy', because female employment in the factory for instance could be regarded as an anomaly. Cultural and religious requirements of feminine modesty make jobs that involve tremendous physical rigours and frequent interaction with non-family male counterparts unsuitable for women. Islamic and Hausa traditional conception of exemplary womanhood includes humbleness, taciturnity, and submissiveness. Conversely, the keen competition, aggressiveness, and intimate struggle with males that factory working conditions foster are thought to rob a woman of these qualities.

The evidence from the Chinese factory revealed that the most outspoken female employees, who mixed freely with their male counterparts on the shop-floor, were treated with euphemism and regarded as wayward, or *karuwai*. Male workers perceived managerial recruitment of female applicants as amoral. Muslim male employers in Kano are usually cautious and reluctant about recruiting female employees because of the cultural and religious implications of such actions. Reacting to the question of female employment in the factory, a local Hausa business industrialist made the following remarks:

> In this factory, I am very cautious about employing female workers. You can see that there is just one non-indigenous female worker in the reception room, who is

> also a married woman. I particularly do not want to recruit single women and be labelled by society as an *Alhaji* providing sanctuary for prostitutes (*karuwai*).

The discrepancy between female and male employment in factories is not peculiar to Kano, but is a general trend in Nigeria and other developing societies. The so-called female subordination is not unique to northern Nigeria, but appears to be universal to all cultures (Callaway, 1987, p. 21). A study by Fapohunda (1976) shows substantial variation in labour force participation between males and females across the country. There is significantly higher male participation than female in the formal economy.[217] Furthermore, a report by Okonjo (1980) on extension workers in the agricultural sector of Bendel State showed that women in agricultural related jobs were marginalised. Low female participation in the formal urban economy is not only restricted to the predominantly Islamised parts of northern Nigeria. It is however, necessary to mention that women's labour participation in the formal economy is higher in the Christian dominated south than in the Muslim North. Women's dual role at home and in the work place sometimes inhibits their capability of competing favourably with their male counterparts. Unlike their male counterparts, women are often less suited for transfers to new destinations because of their traditional role as mothers and custodians of the home.

In the more recent times, there has been a bifurcation of the Nigerian labour market along gender lines. This scenario has created misleading notions and bipolar identities of 'productive males' and 'non-productive females'. Employers of labour are particularly less willing to employ female workers. In factories where women are employed, business owners are often reluctant to allow them to take maternity leave or continue to pay their salaries and allowances in abstention. Because of their vulnerability to sexual harassment, the Nigerian Labour Decree of 1974 prohibits the use of female workers on night shifts, except those in the medical, paramedical, and management fields. The same decree, however, gives a woman the right to leave her place of work for maternity leave once she produces a letter from a competent and registered medical practitioner confirming her pregnancy and child expectancy in six weeks. These and other legal matters binding female employment further increase the entrepreneurial and managerial preference for male employees in the factories.

The general pattern of few women workers in factories became apparent during my fieldwork in Kano. Some of the factors responsible for this may include the low level of female education in the community and cultural definitions about wage employment.[218] Female factory workers made up only a small fraction, approximately eight percent, of the entire population of the labour force in the three factories. All the female employees except one came from the Chinese factory. The Chinese factory was also the only case with a female Filipino manager. It is, however, unclear about whether the Chinese management was favourably disposed towards female employees or felt compelled to hire them to do the simple repetitive task for which females were supposed to be best suited.

A cross-tabulated profile of the female workers further reveals that most of them representing approximately 81 percent were Christians and hailed from the middle-belt and southern parts of the country. The minority Muslim female factory workers, who represented 19 percent of the population, were mainly of Kanuri and Babur ethnic backgrounds. Although the oldest female in the factories was 45 years of age, most of the women employees were within the age category of between 31 and 40 years. This contrasts sharply with the much younger male workers, about 52 percent of whom were between the age category of 21 and 30 years. Approximately 50 percent of the female workers in the Chinese factory were married, while 43 percent were single or unmarried. A relatively small number, constituting seven percent were divorced or had experienced life between marriages. However, like their male counterparts, the educational qualifications of most female factory workers were very low, often not extending beyond the secondary, primary, or Quranic school level.

Puritanical Islam and the Economic Woman

The general framework as outlined in this study suggests low female participation in the formal industrial sector of Kano. The 1991 population figures from the Census Office in Nigeria showed that females made up 2,773, 316 out of the total population of 5,632,040 in Kano State.[219] A tacit implication is that, in a society where women make up nearly half the population, it may seem that their low or non participation in the open or formal economy will lead to general stagnation and economic decline. An observer, of the Kano economic sector over the years will notice an appreciable growth at least in the commercial sector.

In recent times, the role of Muslim women in paid employed has aroused a significant amount of controversy in northern Nigeria and parts of the Islamic world. That Muslim scholars and many Hausa adult males are unanimously opposed to female employment in the formal economy does not mean that they are against female productive economic activities in the households. Muslim scholars insist that the ideal places of employment for Muslim women are in teaching, nursing, and medical care, which often require some considerable amount of intimacy and feminine privacy. Drawing from the negative consequences of the so-called women liberation or emancipation in Europe and the United States, Al-Bar (1984) contends that this has only created social and psychological problems related to corruption and sexual promiscuity. Today, there is a crisis of the family in the Western world. This crisis has resulted in higher out of wedlock births and increased divorce rates among working than non-working mothers.

Female employment in a mixed environment is generally condemned by many Islamic scholars. As such, female employment may be acceptable where it does not infringe on religious ethics and norms. This implies, female's paid employment, must be under separate conditions with other male counterparts. In Muslim countries such as Saudi Arabia, Indonesia, and Malaysia, there are separate working places and banks exclusively for women. It is pertinent to mention that such provisions for Muslim women are not available in any part of Nigeria.

Puritanical or pure Islam combines simplicity with strict orthodoxy. In its ideological manifestations, it deals with the total submission to Allah and adherence of the teachings of the Qur-an, Hadith and the *sunnah* or traditions of the Prophet. Submission to the will of God encompasses all human activity including spiritual, political, social, and economic. Thus, Islamic economic rationality is based on a simple philosophy that economic pursuit must be consonant with religious norms. A fundamental principle of Islamic puritanism is the notion that a woman could keep religious rules, work within the confines of the home and still be very productive to society. The tremendous scholarly contributions of Aisha the wife of the Prophet, to the Islamic world were conducted at home. It is quite fascinating to note that over 1,200 *Hadith* were transmitted from her. Similarly, in northern Nigeria, the Shehu's daughters, such as Khadija and Nana Asma'u, are known for scholarly contributions in the religion of Islam. It is significant to mention that, all the scholarly activities of these women were conducted at home.

Muslim women have full control over their capital resources,[220] and are therefore not disadvantaged in their economic pursuits by seclusion or *kulle*. The history of Islam shows that Khadija, the first wife of the Prophet was a prosperous woman who operated 'her own caravan business in Mecca in the seventh century' (AlMunjjed, 1997, p. 82). Prosperous Muslim women in the early years of Islam often delegated responsibilities to males on economic matters that required extensive travels and absence from home and intimate interaction with other non-family male counterparts. In spite of this, women can be sole trustees of their businesses without the need for intermediaries. Muslim, women have made tremendous contributions to their societies since time immemorial. In the early years of Islam, women accompanied Muslim armies, 'nursed the wounded and prepared supplies for warriors' (AlMunajjed, 1997, p. 20).[221] Their relative dependance on males apart, Schildcrout (1983) observes that Muslim women exercise a significant amount of autonomy in their daily activities.

Islamic and Hausa cultural ethics discourage idleness. This implies that 'a woman in her child bearing years without a trade is an anomaly, so too an old woman without her own financial resources is an object of pity' (Coles and Mack, 1991, p. 125). Society is not only aware, but positively encourages the productive economic activities of women in the households. Successful merchants often spawn the household trading activities of spouses and other female members of the family. Female quest for personal income further encourages petty trading and other craft activities in the households.

Coles (1991, pp. 163–191) study in the northern Nigerian city of Kaduna shows that about 80 percent of the Hausa women interviewed had an occupation or *sana'a*. Nearly half of the women interviewed had multiple income generating activities which include the 'sale of fried groundnut cakes *kuli kuli*, noodle *taliya*, spices *kayan miya*, mat *tabarma* cap or *hula* weaving'. Many secluded women who engage in petty soap making and small-scale plastic bag manufacturing in areas such as Kundila occasionally hire male workers to help them in their productive activities. Today, secluded Hausa women in Kano double as commuter transporter owners. The economic activities of secluded women are highly labour-intensive. Male intermediaries often act on behalf of women, as bus drivers or motor bike riders popularly known as the *Yan acaba*.

In analysing the hidden economic activities of Muslim Hausa women, Schildkrout (1986, p. 195) observes that child hawkers[222] are significant economic actors. As much as children may enhance adult female economic activities, they are sometimes 'entrepreneurs in their own right' (ibid.). In particular, secluded women helped by child hawkers are found to have higher incomes than those do not use their services (Schildkrout, 1979, and 1981). Through the assistance of children, many secluded women in northern Nigeria sometimes contribute between 20–50 percent household subsistence.

Strategies for commercial capital formation among female business groups include personal savings and through the rotating credit scheme or *adashe* discussed in the Second Chapter of this book. Normally, a more elderly and responsible woman in the community or household heads the pool as *uwar adashe*. Credit facilities are disbursed at a time deemed appropriate or when a member was in critical need of commercial capital. This implies that sometimes, the rotation rules may be bypassed to lend support to the most financially handicapped among the group. Nevertheless, the non-recognition of hidden economic activities by the modern financial institutions implies that female access to loans or venture capital is very low. In view of this, very few Muslim Hausa women seek credit facilities from modern financial institutions.

Yet, personal income is crucial to the Hausa woman, in the fulfillment of social obligations to her daughters and those under her care (Coles and Mack, 1991, p. 124). Such social obligations include accumulation of dowry property for young girls awaiting marriage. Dowry property consist of boxes of wedding dresses, wooden cupboards, enamelled pots and pans of different sizes, clothes, trunks, luxuriously decorated beds etc., which require heavy capital investments from lifetime savings. Writing on Hausa architecture, Moughtin (1986, p. 138) observes that women who are responsible for inner decoration of compounds rooms often spend much of their incomes on this. Today the challenges of urbanisation, economic hardships, and spiralling inflation in the Nigerian economy have increased the need for female household trading and financial independence.

Female Activities in a Hidden Economy

Since men's success is incomplete without the contributions of woman, it is justifiable to assert that the economic prosperity of the merchant classes in Kano and parts of northern Nigeria was made possible through the hidden activities of secluded women. Secluded homemakers often relieve men of a greater burden of household chores by giving them enough time for their economic activities. Domestic unremunerated female services are strongly intertwined with male income earning activities.

Peasant cash crop production of agricultural export commodities in the colonial economy was heavily subsidised by the labour of poorly paid rural women. The role of Muslim Hausa women in agriculture, cultivation, harvesting, processing food for household and for local consumption cannot be ignored. Under the commodity based export economy, rural women played significant roles in pounding, threshing, cleaning, and bagging export products. As such an analysis on the economic prominence of Kano as centre of peasant export crops would be incomplete without stating the contributions of women especially in the rural areas. Such contributions are often glossed over or not properly documented by historians in patriarchal societies.

Activities of women in the economy depend on whether such an economy is shrinking or expanding (Frishmann, 1991). Between 1993 and 1997 the period within which this study was conducted, the commercial sector has witnessed significant structural face-lift. The Abubakar Rimi central market and the Fagge commercial centre in which several hundred thousand traders conduct their economic activities have been transformed to modern business complexes. New commercial areas have also emerged in the Hausawa quarters. The expansion of the Kano business sector has opened income-generating avenues and boosted female economic activities in the households. The degree of economic activity among secluded women varies among communities and households.

Through petty household trading, secluded women are sometimes able to accumulate substantial capital through shrewd spending. In urban centres, economically successful women sometimes embark upon *hajj* or sponsor their spouses' pilgrimage trips to Mecca. Such trips increase women's social prestige. Women's economic earning forms a substantial part of the subsistence need of the family. Women make hidden socioeconomic contributions as wives, mothers, and guardians of households, which are often unremunerated. If the free social services of housewives were taken away and transformed into paid services, the economic cost will be too substantial and beyond the capability of most spouses. For Glucksmann (1995, 70), 'the economic bottomline' of women's contributions is 'that if babies are not looked after they die; if food preparation ceased people would eventually starve'.

The economic success of merchants implies that the household consumption pattern of Muslim Hausa women in Kano is significantly high. Even as consumers of goods and services in the society, secluded women make a direct positive impact on economy. When considered from a productivity angle, increasing consumption patterns and demand for goods and services fuels high economic productivity. Conversely, productivity is small were consumption patterns are low. In vibrant economies, household economic demands are crucial in propelling the production of goods and services. Urban household demands, for gowns, dresses, and ritual rams keep the wheel of the northern Nigerian economy in motion. Through hidden activities in the informal sector, secluded women can bypass not just the formal economy, but even the markets.

Although seclusion may influence the form and nature of women's participation, it does not cut them off from political, economic, religious and artistic aspect of life (Coles and Mack, 19991:12). Conversely, seclusion does not imply that Muslim women are passive consumers or have no choice to determine the course of their development process. To a large extent, they do. Like their male counterparts, most secluded Hausa women pursue their economic activities with an air of assiduity. Motherhood and domestic chores have not diminished the economic activities of these women.

Their adherence to seclusion norms notwithstanding, Hausa women in especially Kano sometimes do have international business connections. Some of their international business trips are seasonal and conducted during Islamic pilgrimage to Mecca. This is significant because it provides avenues for spiritual exaltation and personal economic opportunities. More recently, Hausa women who have substantial capital base are actively engaged in business with predominantly Muslim countries such as Saudi Arabia, Dubai, and Egypt. Mecca is a common destination for Muslim businesswomen who travel. Modernisation implies that an affluent Hausa woman with communication facilities at home such as the telephone may transact business with other women without having the need to move beyond the house. Economic transactions across the telephone though prevalent among urban Muslim women of aristocratic and merchant classes are uncommon among rural Hausa women.

The operation of direct air shuttle services by local airlines between Kano, Jeddah, and Dubai has further boosted female foreign business travels to such Muslim countries. The evidence suggests that most of the urban upper class Hausa women who engage in such business activities are divorced or have experienced life between marriages. In spite of this, female business travels to Western or non-Muslim countries can be viewed with suspicion or euphemism. Most Hausa business women use the Muslim pilgrimage to bolster economic transaction. Women's international business activities are annual or biannual in nature and often take a form of *hajj* and the lesser-*hajj* or *umra*. Sometimes, a religious visit to the Holy Land in Mecca can also be used as medium for business transaction. Business merchandise for urban business women often includes jewelry, clothing, and cosmetics. Hausa business women know their customers and market niches among the prosperous women in purdah who are less privileged to travel.

Nonetheless, the contributions of women in economic development are frequently overlooked by studies of Islamic societies. Many hidden economic activities of women go unnoticed or unrecorded and excluded from economic analysis. In the informal sector, the productive activities of traditional midwives, herbalists, weavers, food sellers, soap makers, and tailors cannot be easily overlooked. Thus, 'women in every society play a crucial role both indirectly, by supporting other workers with household activities and directly, by participating in economic production of goods and services for sale' (Frishman, 1991, p. 192). Domestic activities have been officially omitted under the United Nations unemployed labour force data (UN, 1975, p. 62). The activities of women in urban economy are neglected or intentionally overlooked by planners and development economists (Frishman, 1991, p. 192). Women's economic activities are sedentary in their outlook because they are conducted at home. The omission of hidden economic activities of women in northern Nigeria and developing countries is therefore a product of defining work and employment within the economics boundary. Economics has become a 'discipline representing the economy' and work was subsumed within its boundary (Glucksmann, 1995, p. 66). In view of this, every non-wage activity or transaction that occurred outside the economy was not analysed as economic (ibid.).

Frishman (1991) observes that the United Nations Economic Commission's report on labour productivity for 19 countries in Africa showed that in manufacturing sector, the median for women was 8 percent. The conclusion according to this report was that 'women play a relatively small part in modern wage sector' (UN, 1975, pp. 62–66). Paradoxically, the report added that only the formal economy, the most easily accessible was measured by economics analysts (Frishman, 1991, p. 193). There is however, ample evidence in Kano and parts of the developing world to show that work can 'occur in any sphere or institution outside the formal market sector'.[223] The productive but hidden economic activities of secluded women in Hausaland are particularly common and they offer substantial contributions to the Nigerian economy. In estimating the value of hidden productive activities, Frishman's (1991, p. 203) data show that the contributions of at least 100,000 women engaged in hidden trade in Kano are comparable to the total industrial wage bill. It is significant to mention that, there are today over half a million Muslim Hausa women in seclusion engaged in productive economic activities. The evidence suggests that the impact of women on the economy is substantial.

This, Chapter shows the economic rationality of a Muslim Hausa society. The analyses suggest that, in Islam, women could still practice religious norms, operate behind the confines of the home or *harem,* and still be economically productive to society. In view of this, it will be fair to assert that neither the adoption of Islam nor female seclusion has inhibited the economic activities of Hausa women. Much of the controversy however, as Glucksmann (1995) has observed, centres on the exclusion of female household activities outside the economic sphere. That secluded Muslim women are able to bypass the open market in their economic activities does not imply they are none-productive.

The evidence suggests that female participation in the hidden informal economic sector is substantial and comparable to the contributions of industrial labour in the northern Nigerian economy. The per capita income of secluded Muslim women is also comparable to those of many males in northern Nigeria. Domestic female activities in the households as the study shows, relieve men of greater economic burden. As such, the household contributions of secluded women continue to play a crucial role in the sustenance of many homes in northern Nigeria, and will perhaps remain so for many years.

10 Entrepreneurs and Business Success

There are today varied conceptions across disciplines about what is business success. In measuring business success, multiple criteria and standards are applied across countries, cultural and religious groups. Even within a specific society or business environment, the criteria by which entrepreneurs in the service and commercial sectors measure success are usually distinct from those used in the manufacturing sector. To say whether a firm is successful or not depends on the specific cultural context and the parameters used in measuring success. This implies that businesses classified as successful under certain environments may be unsuccessful in another.

Societal perceptions and definitions about business or economic success further reinforce entrepreneurial attitudes about the nature, size, and number of businesses they aspire to operate at a given time. Individual and group definitions about economic success affect general macroeconomic performance and development. Business success is transient and erratic. In business circles, firms that are sometimes written off at the start often become profitable and successful in succeeding years. Since successful businesses are sometimes the products of several failures, it is often analytically misleading to label a firm as a complete failure or success, while it is still in new field of economic activity. This implies that any meaningful analysis of business success should be seen within a long-term frame.

In self-operated businesses, entrepreneurs must confront the difficulty of assessing the success or failure of their economic activities. Sometimes cultural values, past experiences, and personal aspirations shape the way entrepreneurs, managers, and workers perceive success. Impoverished employees usually tend to believe that whoever pays their wages and salaries, must be economically successful. Employee views about business organisations are often at best parallel with those of managers and do not sometimes reflect the true performance of business organisations. From a wider perspective, societies tend to define success within the context of relevant experiences drawn from history. Societies that have never experienced mass production may define small-scale activities as highly successful. It is therefore not strange that, in many developing societies, petty traders and food hawkers are labelled as successful business persons and women. These perceptions and definitions about business success tend to hinge on the micro-environmental context in which the businesses operate.

Business success is sometimes conceptualised within the context of organisational goals and the extent to which it meets or deviates from such goals or targets. In this regard, a firm can be judged successful as far as it achieves its targets and goals, no matter how minuscule. A small percentage of our entrepreneurial and managerial respondents, representing 11 percent, perceived the ability to meet personal goals as a mark of economic success. The self-assessment standard for measuring business success is sometimes misleading for two simple reasons. First, in every society, most entrepreneurs and business operators never admit success, regardless of profit margins and abilities to meet or exceed personal goals. Such business operators do not fit into the self-assessment model. In many Nigerian societies, including the Hausa, it is rare or unusual for individuals or groups to make open claims about economic achievements. Indigenous Hausa entrepreneurs are usually cautious about making open claims about their economic achievements because such utterances and boastful attitudes transcend the ethics of cultural and religious modesty. This, however, does not foreclose the existence of a few eccentric individuals in societies who violate this norm to boost personal egos. To a large degree, entrepreneurial concealment of business operations makes the self-assessment approach weak.

In northern Nigeria, as in many parts of the world, business success is synonymous with profit. An analytical discourse of business success therefore has to have a link with the subject of profit discussed in the Sixth Chapter of this study. In the industrially advanced world, business success is usually measured in monetary and growth terms, whereas, in the Hausa society, business success sometimes transgresses monetary and growth boundaries. Wealth and profit acquisitions, no matter how substantial, may not be seen as a mark of success if they originate from questionable sources. Conversely, economic profits and business growth do not necessarily imply success if they do not have religious sanctity. Entrepreneurial sanctification of business profits may include: the fixing of acceptable prices on goods and services, providing support to family and primordial groups, paying the annual tithe, or *zakkat*, and engaging in charitable activities for the sake of Allah.[224]

Success in Islam is synonymous with spiritual exaltation or self-purification.[225] Entrepreneurs who make huge profits but show little concern for social and religious obligations may be labelled as complete failures in Hausa society. In contrast, a business person who earns little, but fears God, has sympathy for the needy, and is a patron, or *uban-gida*, of less fortunate members of society, may be seen as more successful. Apparently, not all successes are considered positive, admired, or even worthwhile. Success is therefore metaphorical and connotes a myriad of things in the society. At the two extreme ends of the continuum, entrepreneurial success may be branded as 'positive' and 'negative'. An indigenous industrialist sums up his understanding of business success as follows:

> Business success is not a straightforward issue. However, it can be explained that there are two distinct types of successes. One is positive and the other is negative. The negative type of success is synonymous with the success of the proverbial well-digger. The success of the well digger deeper down the bowels of the earth, implies separation from the rest of humanity . . . The progress of the well-digger, while sinking the earth has an inverse relationship in terms of separation with rest of the society . . . You can see that no sane person wants this type of success or progress. Positive success, on the other hand, is the type of success which no matter how trivial, brings a person closer to his creator Allah and the community in which he lives.

The view expressed by the local industrialist was universal among both indigenous workers and their employers in both the commercial and the industrial sector of Kano. Since the definition of success goes beyond economic achievements, what a businessperson does with his/her wealth in his relationship with God and Society is highly important. Entrepreneurial management and use of wealth eventually define his/her position in the society. In the commercial sector, the use of influential networks of contact further enhances economic success.

Table 10.1 represents entrepreneurial perceptions about business profit in the business community. The responses have been categorised under five main indices to explicate entrepreneurial views about business success. Approximately 59 percent of the respondents, who were mainly entrepreneurs and mangers of business firms, perceived the ability to earn and use profit in a socially accepted way as a sign of business success. Approximately 18 percent of them saw growth and expansion of business as the most befitting characterisation of business success. The link between growth and success is common in nearly every society. The bigger the firm, the more successful it is perceived to be. Sometimes small businesses are erroneously seen as failed or unsuccessful businesses. Associating business size with success can sometimes be very misleading because of the fact that very large firms tend to incur higher overhead costs on services such as electricity, and water, which often overwhelm business profits (Harsh, 1986, pp. 72–76).

Table 10.1. Entrepreneurial/Managerial Perceptions of Business Success

Value label	Frequency	Valid Percentage
1. Success as the ability to earn and make efficient use of profit	26	59.1%
2. Business success as growth or expansion	8	18.2%
3. Business success as absence of debt	2	4.5%
4. Business success as the ability to meet personal goals and targets	5	11.4%
5. Business success as continuity in operation	3	6.8%
Total	44	100%

Source: Author's field interviews, 1995.

A rather small percentage, forming approximately 5 percent, perceived business success as absence of debt. Entrepreneurs with such narrow definitions of economic success seldom aspire to large-scale expansion, due to their aversion for debt. In many developing societies, people sometimes see the operation of diversified or multiple businesses as a sign of success. Persons with diversified business interests are perceived as having a wider capital base and a capacity for handling different things at a given time. In the local community, sometimes the size and structure of housing and number of dependents are the most distinguishing signs between the affluent and the poor.

As with profit and size of a business firm, many years of operation or perpetuity in business operation may be perceived as a sign of success. This stems from the widely held notion that only failed businesses disappear. Few respondents, making up approximately 7 percent, saw continuity in business as a sign of success. This definition perhaps is derived from the fact that most people will stop pursuing an economic activity if it proves to be unsuccessful.

Approximately 57 percent of the entrepreneurs and managers saw the failure of their previous businesses as reason for leaving or diverting into new economic fields. The notion that perpetuity in business is a sign of success also has major limitations. In most African communities, entrepreneurial endurance in specific non-prosperous economic activities is sometimes an attempt to avoid the stigma of being labelled as a failure or idler (Tahir, 1975). In many craft occupations, sometimes the reasons for continuity are borne out of the desire of artisans to protect 'family secrets' from disappearing.

In measuring business success, independent observers provide very relevant insights about the performance of a firm. The independent observers, in this context, refer to the customers and business associates of a firm. In such instances, the historical accounts of customers, which are derived from past and present interactions with the business firm, become the basis of assessment.

Because of limited resources, this study has not been able to incorporate the views of customers regarding the success or failure of the business firms being studied. Nevertheless, it is significant to mention that such assessment can provide useful guides for the analysis of business success. However, in spite of the prevalence of other external variables, the evidence presented in this study has shown that cultural and religious values tend to exert a great deal of influence on entrepreneurial perceptions about business success.

Business Success: An Elusive Concept

Due to the general lack of a consensus at both the theoretical and the practical levels about what makes up business success, there is a widespread acceptance about the elusiveness of the term in recent years. Business success has been interpreted in terms of several variables such as the rate of return on capital and sales (Harper, 1987). For Beveridge and Oberschall (1979), the most important indicator of success is the financial turnover. Although there is a correlation between financial turnover and business success, this also has some analytical handicaps.

Profits returns on capital are not sufficient measures of business success, for the fact that accumulated debts may sometimes exceed profit margins. Furthermore, the business turnover assessment places emphasis on sales and ignores depreciation costs. However, neither growth nor the ability to exceed targets is a satisfactory indicator of business success. Associating business success with perpetuity can also provide false indicators to business performance. The mere fact that a firm continues to operate often does not imply that things are moving smoothly. Today, the application of multiple standards in measuring business success further makes the idea elusive. Since definitions of business success also rest on the parameters or criteria that are used, business success therefore lies in the eyes of the beholder.

A study by Wooddruff and Alexander (1958) shows that successful businesses are characterised by good financial records and systematic decision procedures by operators. Similarly, Baumback's (1983) study identifies four conditions for business success which include, alertness to change, the ability to adjust, the ability to control operational details and a good forecast of knowledge of the market. Differing interpretations and parameters for measuring success notwithstanding, the *sine qua non* conditions for successful enterprise are sustainable profits, growth, and perpetuity. On perpetuity or experience, Liedholm and Chuta (1976) observe a positive correlation between the age of a business and profitability. Conversely, debts, shrinkage, and closure of business often become the yardsticks for measuring failure.

On the basis of the first three parameters, the Hausa and Lebanese plastic factories have been classified as moderately successful, while the Chinese firm, which scored highest on these parameters, is ranked as very successful by local standards. It should also be mentioned that because of the impressive performance in their various manufacturing activities, the proprietors of the three plastic firms, were recipients of Geneva International Awards for excellent performance and for quality products at the various times of their operations.[226]

In Kano, as in many African societies, the sources of business success are not skill, knowledge, and self-confidence, but are essentially derived from luck. The idea of luck, or *sa'a,* is central to the African business philosophy and is universal to all societies. Even in the industrialised world, luck is sometimes counted as an essential component of business progress. However, unlike the developing societies, the idea of luck is epiphenomenal in pragmatic economic pursuits. In the semi-industrial society of Kano, luck is perceived as a significant factor on which business success is based. A daily Hausa aphorism that 'business is luck' sums it all. The pervasive notion of luck in business sometimes creates nonchalant attitudes and complacency among business operators once moderate achievements have been made.

Furthermore, the indigenous entrepreneurs' strong belief in luck as the ultimate source of business success often leads to poor planning and, sometimes, the failure of their businesses. Among Hausa entrepreneurs, examples dealing with luck and wealth acquisition are sometimes cited from religious texts. It should be stated that, although the Qur-an talks about luck, it nevertheless emphasises perseverance and hard work as the basic requirements for success in any human endeavour. In contrast to the sacred texts of other religions which 'discourage economic activity in general, counselling their followers to rely on God to provide them with their daily bread', the Qur-an encourages the pursuit of economic activities at all times including *hajj* (Rodinson, 1974, p. 14). The link between Islam and economic pursuit is therefore a direct one. A fundamental principle of Islam is that 'God will never change the grace which He has bestowed on people until they change what is in their souls'.[227] This implies that practical self-help and hard work are the essential qualifications for divine assistance, as opposed to reliance on luck. In Islam, if a man is unsuccessful in a particular trade, he is encouraged to try another. Imam Ghazali (1961; [1058–1111]) stresses perseverance and change of strategy as means of attracting customers and boosting incomes. Consequently, a combination of entrepreneurial openness, perseverance, and low pricing becomes the market penetration strategy.

Schumpeter (1962) postulates that 'supernormal intelligence and energy account for industrial success and for the founding of industrial positions in nine cases out of ten'. This implies that although 'nature is cornucopian', yet 'it is not freely available, it is controlled by skill' (Wildavsky, 1992, p. 157). However, the supernormal intelligence and skill that Schumpeter advocates for successful entrepreneurship are the exceptional innate human traits. An observer of business activities in Nigeria, and especially Kano, will notice that the majority the most economically successful businesspeople are people with extraordinary zeal in their economic dealings.

Sometimes poorly educated, but hardworking entrepreneurs have flourished in economic activities where their well-educated colleagues have failed. By these achievements, the merchants of Kano have proven that educational handicaps do not mitigate the zeal towards economic success, especially in small-scale business activities. Harper's (1984, p. 14), study of successful Kenyan businesses run by uneducated indigenous persons offers some caution about generalisations. He observes that while uneducated persons may be successful in operating small businesses, their abilities to manage large-scale enterprises effectively are sometimes questionable.

Why Most Businesses Fail

The histories of most developing communities are replete with accounts of business failure. In spite of the pervasiveness of this scenario, the statistics of failed businesses are difficult to obtain in most developing countries, including Nigeria. These difficulties stem from a number of reasons. First, Nigeria is a mixed economy in which the informal sector operates quite independently of the formal economy. Poor official monitoring and regulation of the informal sector have resulted in a paucity of information on a large volume of small business activities in the most dynamic sector of the Nigerian economy.

Second, the secret nature in which businesses are conducted in the informal sector often means that sometimes public officials, spouses, children, and relatives are kept entirely ignorant of diversified business interests.[228] Added to this is the fact that business operators do not openly admit failure or financial bankruptcy for fear that it may be interpreted as admitting incompetence, or inefficiency. Entrepreneurial reluctance towards admitting failure sometimes has to do with fear of being perceived as greedy or ungrateful to Allah. Usually, business operators make moderate claims about business performance as a way of avoiding societal stigmas.[229] Official sources from the Nigerian Manufacturers' Association (MAN, 1995) show that the rate of failure, of indigenous manufacturing ventures, is very high. It is necessary to state, however, that this high rate of failed businesses is not peculiar to Kano or northern Nigeria, but is a general trend even in the industrialised world. A study of businesses across Europe and America reveals that:

> Out of every one thousand (1000) new businesses started during the year, two hundred and fifty (250) disappear before the end of the year. By the end of five (5) years, only three hundred and fifty (350) remain; and no more than one hundred and ninety (190) of the original start their eleventh (11th) year. These figures apply to equally to all types of businesses (Poughkeepsie, 1969, p. 13).

A study on business failure records by Baumback (1983) suggests that 'poor management, insufficient capital and economic depression' were among the factors listed by creditors and ex-business owners. In Nigeria, several factors often conduce to make businesses fail. The factors can be categorised as endogenous and exogenous. In exploring the two factors responsible for business failure, emphasis is placed on the former. Exogenous factors include bureaucratic bottleneck, policy instability, and poor infra-structural facilities such as poor road networks, erratic power failures and inadequate supply of potable water. Frequent macro-policy changes often imply that long-term planning in business activities is impossible. Unnecessary bureaucracy in policy pronouncements and their sloppy implementation have often resulted in immeasurable losses to business operators in Nigeria. One respondent, an indigenous industrialist, bitterly complained about his loss of two million naira during a single month in 1994, because of policy change concerning foreign exchange allocation.

Political instability and frequent policy changes sometimes stifle the capital of small businesses in developing countries. During the first quarter of 1995, official delay in the announcement of the annual national budget was largely responsible for heavy congestion of goods at the Nigerian ports. Bureaucratic delays and their resultant consequences sometimes compel manufacturers and importers to wait for months for the release of new policy guidelines before goods can be cleared from congested sea ports. Very often, the time lag between the receipt of raw materials and spare parts and their utilisation in the manufacturing process is too long and unsuitable for manufacturing activity, which hinges on continuity. This scenario usually leads to low capacity utilisation in most firms in the manufacturing sector. The perennial shortfalls in production targets lead to quality variation, customer dissatisfaction, low patronage, loss of profits and the failure of most indigenous manufacturing firms. Although expatriate firms are not spared these ordeals, their international connections, control of extensive capital, and the ability to airlift raw materials directly from overseas without having to pass through congested national seaports, puts them at an advantageous position over their indigenous business counterparts.

Demand limitations

For a business to arise there must be demand for its services. The growth of factories in Kano over the years has been due to increasing demand for industrial goods. The buyers of made-in-Nigeria products are usually poor people such as peasants and the jobless. Although not necessarily of low quality, inconsistencies in production standards often give local goods the image of poor quality, therefore their demand among the urban middle and upper class is low (Harper, 1984, pp. 120–122). With low incomes, mounting inflation, and increasing impoverishment in the local communities, there is now a decline in the patronage of domestically produced goods. The low demand for home-made-goods is a major bottleneck for business expansion.

Unlike big multinational firms, small business enterprises often do not have sufficient resources and knowledge to promote or advertise goods through the mass media. The circulation of indigenous products is therefore limited to the local environment in which they are produced. Such businesses are also handicapped by inadequate transport facilities, which implies that the goods that they produce are poorly distributed. This sometimes leads to over-competition and gluts in the markets in which the factories operate, while there may be scarcities in the country or region. It is also becoming clear that societal demands for goods and services are conditioned by the societal standard of living. The crux of the matter in the current low demand for industrial goods is the poor economic climate in Nigeria.

Managerial constraints

A basic argument that is often made about the poor performance of Nigerians in business endeavours is the 'managerial and technological deficiency' thesis. Kilby (1969) is of the view that the absence of a major breakthrough in Nigeria's industrial sector is the result of managerial and technological deficiencies. According to him, the managerial strategies employed by Nigerians in the industrial sector are inadequate and unsuitable. Although such a classification is an over-generalisation because of variations in managerial ability and technique across Nigerian groups. Some inhibitions to successful local business management include entrepreneurial access to information. In particular, an awareness of market potential or the rate of business failure in a particular environment, which is crucial for business, is often lacking. This implies that regardless of the zeal to pursue business, lack of relevant training and information on sources of raw materials, durable machines, and technique of measuring employee productivity can be major reasons for poor performance. Sometimes due to lack of information on certain business activities, prospective indigenous investors see them as attractive, while in real practice they are not.

A typical characteristic of indigenous entrepreneurs, especially in the commercial sector, is their vulnerability to the bandwagon effect. Imitation of others, without a knowledge of the capacity and strength of the market, can be detrimental to business performance. Many indigenous entrepreneurs are hardly aware of the potentials, limitations, and seasonal fluctuations of the market in which they operate. This development has often led to the failure of new entrants in business, especially in Kano. Apart from access to information, the psychological preparedness of the business entrepreneur is very important. Most businesses fail because of inadequate preparedness. Sometimes entrepreneurs only discover the enormity of the task they have chosen to pursue when it is impossible to withdraw. This has been the situation with some industrialists in Kano.[230] Strenuous requirements of business activity may include working longer hours sometimes solving unforeseen problems, or embarking upon unexpected journeys during certain odd hours of the day, week, or month. Since profit margins are never uniform, incomes in business activity are also hardly regular. The truism that incomes are seldom regular in business is a fact that most business operators in developing countries find difficult to swallow. An adaptive response is therefore multiple investments to ensure a regular flow of income.

A study of failed businesses by Duvel (1962) reveals that poor financial records are among the major causes of business failure. However, studies by Marris and Somerset (1971) on indigenous businesses in developing countries have debunked the poor bookkeeping claim as baseless. The evidence from this study suggests that in spite of the poor maintenance of financial records in the plastic firms, they were successful in their economic pursuits. This perhaps shows that, as far as developing countries are concerned, the poor maintenance of financial records might not be a strong factor for business failure.

A considerable amount of literature in business management underscores the role of entrepreneurial commitment to marketing as a major determinant of business success (Cohen and Reddick, 1981). This is because marketing embraces a whole gamut of activities that include the coordination of all productive effort from the beginning until the product reaches the consumer. Businesses that do not market properly normally flounder. Indigenous operators of small firms often feel that marketing is only a concern of big business which is a mistake because big businesses often survive marketing mistakes than smaller ones (ibid.). None of the three plastic manufacturing firms had a marketing department. This implied that most of the activities involving marketing were left either undone or attended to haphazardly.

Employee work habits

As with entrepreneurs and managers, the literature on the performance of Nigerian workers in industry or under wage employment finds their performance less than impressive.[231] According to Berg (1968, p. 400), the lackluster performance of Nigerian workers in industry is a recurrent phenomenon in other traditionally oriented or preindustrial societies. In contrast to the findings of Berg and others, studies by Kilby (1961), Harris (1968) and Hogendorn (1978), on the conventional appraisal of Nigerian workers' productivity have debunked the negative labels as totally fallacious. It is necessary to state, however, that values are not static; they can change over time.

In every business organisation, employee motivation has a direct impact on general performance. A highly motivated labour force can improve the fortunes of any firm. Poor employee motivation can lead to low productivity and the collapse of a business enterprise. In spite of poor motivation, many Nigerian workers tend to feel that expatriate business owners are less tolerant of subordinates than their local counterparts. The fear of dismissal in expatriate firms encourages compliance with managerial rules, which ensure the meeting of goals and targets in expatriate firms. Furthermore, such fears encourage higher productivity, profits, and the attainment of success in expatriate businesses.

Well over 80 percent of the employees in the Chinese firm were of the view that Nigerian factory workers are more tolerant with expatriate business proprietors than with their fellow-countrymen. Most employees were of the opinion that the twelve daily working hours in the Chinese firm would never have been tolerated if they were imposed by an indigenous entrepreneur. In contrast, managerial permissiveness in the Hausa plastic firm resulted in employee sluggishness and non-compliant attitudes. These attitudes were perhaps derived from the feelings among workers in the Hausa firm that, should things go wrong, the intervention of influential persons could steer the course of harsh disciplinary measures and make the retention of jobs possible.

Cultural attributes

Apart from the technical factors herein elaborated, there are in semi-industrial societies cultural attributes to business failure. Although entrepreneurial inefficiencies or recklessness in handling money or corporate resources may lead to bankruptcy, the failure of a joint or partnership venture may sometimes be attributed to the contagious bad luck of others. The perception that luck and ill-fate are contagious, limits entrepreneurial enthusiasm about joint ventures. Sometimes the evil machinations of business competitors can be seen as the cause of the loss of capital, or *jari*. Such beliefs are often inimical to the development of modern business enterprises.

However, in view of the Islamic influence and the belief in destiny, or *kaddara*, societal attribution of economic failures to magic forces, witchcraft and evil machinations of others, are now less significant. Today, such beliefs are associated with syncretists and nominal Muslims. Nevertheless, the family factor often serves to threaten the performance of micro-businesses. Sometimes, the use of family members without relevant skills or credentials to fill certain posts is a reason for the poor management indigenous firms.

Symptoms of Business Failure

An empirical discourse on business success is often regarded as incomplete if it offers no links to business failure. While it is easier to identify a failed business because of discontinuity, it is sometimes difficult to tell which firms are successful in most societies. The relevance of a discourse on business failure is that business firms that do not exhibit serious symptoms of failure can be seen as successful. One of the key symptoms of business failure is a deteriorating working capital. In mercantile networks, the deterioration of working capital sometimes means that a business entrepreneur is not able to make direct payments for goods and services. Because of the common practice of advancing goods before financial settlements, the deterioration of working capital is not easily noticeable in the commercial sector. By contrast, deterioration of working capital becomes easily noticeable in the modern industry if wages and current financial obligations are not met. In the manufacturing sector, deteriorating capital may lead to shortfalls in the importation or purchasing of raw materials and retrenchment of workers. The inadequacies of working capital and high debt ratios are seen as primary signs of business failure. Bankruptcy, on the other hand, may occur when a firm's total assets are not able to meet its total liabilities.

An early symptom of business decline in the industrial sector is low productivity. The factors responsible for low productivity include poor supervision and the absence of raw materials. In Nigeria, the perennial scarcity of raw materials due to the shortage of foreign exchange has been observed as the major cause for low capacity utilisation and declines in productivity. This trend, which leads to shortfalls in production and the intermittent closure of small and medium-scale firms, often gives a false impression about business failure in Nigeria.

In business analysis, sometimes a decline in the sales of a firm's product may be seen as a symptom of poor performance. However, this is problematic because sales figures are very dynamic and highly unpredictable. Consequently, short-term sales figures cannot be used as adequate measures for business failure. Nevertheless, there is often a correlation between low sales and declining profits. The failure of firms to make profits, often implies that debts are incurred through attempts to honour obligations such as the payment of employee salaries, electricity, water, and telephone bills etc. Financial obligations sometimes take a negative toll on business savings, and therefore additional debts have to be sought elsewhere. In modern business transactions, nearly all firms take loans and are therefore liable to indebtedness. However, when the debt ratio in a firm is higher than anticipated earnings, the firm can be said to be having symptoms of business failure.

Mass retrenchment of employees in firms is sometimes seen as a symptom of business failure. Since most business operators do not continue to retain workers in the face of dwindling profits, retrenchment usually accompanies declining profits.

An ultimate sign of business failure in a society is the preponderance of abandoned business buildings. Kano does not have many of these. However, in recent times, decaying warehouses and factory buildings with rusty gates could be seen in the industrial locations of Fagge, Bompai, Sharada and Challawa. A profile of the abandoned factories and business buildings in Kano revealed that well over 70 percent of them belonged to Nigerian and Lebanese investors. The prevalence of failed businesses has several implications for society. First, business failure involves a multiplicity of costs to operators, employees, their dependants, and society in general. When businesses fail, negative shock waves are experienced in a community, country, or region. Job losses create social upheavals and family instability.

The failure of businesses also results in a decline of public revenue, a slowdown in development projects, and loss of potential investments from abroad. Usually, prospective foreign and local investors do not invest in a system until such a system portrays signs of economic buoyancy. A common indicator of economic performance in a society is the prosperity of the indigenous sons of the area. The prosperity of the indigenous merchants of Kano was a major factor that contributed to tremendous foreign capital inflow into the city in the early 1970s and up to the mid-1980s. Although Kano still maintains her leading role as a focal point of economic activities in northern Nigeria, the political instability of the 1990s and the imposition of economic sanctions by the international business community has resulted in a decline in foreign investment in recent years.

Some Effects of Modern Industry

It may sound absurd to discuss the effects or implications of modern industry in a developing society where there is an insatiable desire for economic development. However, such analyses are relevant for business operators and policy planners in rapidly growing semi-industrial cities such as Kano, Lagos, and Kaduna where industrial growth is threatening to become a menace to environmental quality.

In modern industrial organisations, success sometimes has accompanying negative side effects. Business growth or expansion has multidimensional implications for a society. When businesses succeed, immeasurable benefits are derived by a society. First, the success of businesses leads to their expansion and sometimes diversification into other economic fields. In most of the changes accompanying economic success, there are often improvements in job opportunities, standards of living, and the quality of goods and services produced in the society.

Provision of employment to idle labour particularly has direct links with changes in the levels of income, improvement in purchasing power and the quality of life for many households. Second, business success often leads to mass production of cheaper and higher quality products. In particular, the economic success of the 'Asian Tigers' has led to their ability to produce cheaper industrial products. The ability to transform the manufacturing industry into a sector of mass production of goods remains a big challenge in Nigeria. Business success is also contagious, in the sense that successful entrepreneurs and firms that are household names more easily attract customers to themselves. Business success has a direct impact on the level of a society's development. When businesses succeed, revenue is generated by government for development projects, which results in the improvement of health and infra-structural services such as roads, clean water, and electricity services.

Nonetheless, the emergence of modern industry in Kano has led to the proletarianisation and commodification of labour in the economy. Entrepreneurial cost saving devices in the manufacturing sector have led to the payment of incredibly low wages and sometimes the replacement of human labour with advanced technology. The importation of advanced technical equipment and machinery in most developing countries where there is abundant supply of cheap labour causes a considerable amount of unemployment. The effects often prove to be detrimental to society.

Evidence from northern Nigeria also suggests that entrepreneurial complacency can make a business firm asphyxiate on its success. Sometimes entrepreneurs in developing countries are wary of success because the exposure it brings in terms of the payment of higher taxes. For some entrepreneurs, therefore, business success is not entirely a complete blessing.

Today, the activities in the manufacturing firms are beginning to take a negative toll on the ecology of Kano and its surrounding districts. Indiscriminate disposal of chemical waste from textile plants, leather tanneries, and paint factories into open drainage and underground water channels, forms a direct threat to water users and the productivity of farmlands around the industrial satellite districts of Bompai, Challawa, and Sharada.[232] The ecological damage caused by industrial wastes has often resulted in clashes between peasants and industrialists. However, peasant preference for compromise over litigation with industrialists often puts them on the losing end. With the expansion of Decree 59 of 1992, and the creation of the Federal Environmental Protection Agency (FEPA), it is expected that the indiscriminate disposal of industrial waste in Kano and other parts of Nigeria by manufacturing and processing firms will be effectively curtailed.

In spite of these limitations, sustainable growth and long-term development of the business sector are desirable conditions for nearly every society. Business success in every society is seen as a catalyst for development. However, while it may be much easier to identify the signs and symptoms of business failure, the prognoses for recovery and success are not easy to provide. Since the failure of businesses has negative costs to societies, the need for government intervention to rescue failed businesses in the less developed countries cannot be overemphasised. This requirement, however, remains a very complex challenge for development planners in Nigeria and in parts of the developing world.

11 Economic Development: A General Appraisal

This Chapter addresses the development question by reviewing contemporary development paradigms, their theoretical relevance and practical application to developmental problems in northern Nigeria. The aim here is not to set up new concepts or theories, but to reconceptualise or reinterpret well-known theories and notions of modernisation from cultural vantage points. Some of these culturally enhancing factors to development were highlighted at the beginning of this study in Chapter One. Nevertheless, dependency and world system perspectives are useful in providing supporting evidence to the current economic crises. I begin the framework of analysis by looking at both development and underdevelopment from a general African context. In the second and third sections of this Chapter, the key obstacles to Nigeria's economic development, which include corruption and business fraud, are examined.

Contemporary debate on Africa's performance in political, social, and economic spheres show that the continent is passing through crises of unprecedented proportions. A considerable amount of evidence also suggests that some basic problems regarding political stability, economic matters and food production have not improved; rather, they are worsening. While the rest of the world is undergoing profound changes in economic development, Africa is still at the fringes. Consequently, a larger part of the continent wallows in poverty, while local efforts continue to be hamstrung by huge debt burdens, unfavourable terms of trade, and reverse resource outflows (Salim, 1992, p. 11). Rodney (1972, p. 29) observes that the paradox of underdevelopment lies, however, in the fact that the world's naturally richest parts are the poorest in living standards.

Scholarly efforts toward reconstructing Africa's developmental puzzle remain intense, although the panaceas for solving the endemic problems are hard to come by. Usually, the developmental theoretical approaches applied in addressing Africa's problems coalesce into two major strands: dependency and modernisation perspectives. The collapse of the colonial development project and the general disrepute into which modernisation and dependency theories have fallen has led to the suggestion of a paradigmatic shift (Hydén, 1983). Some limitations of these theoretical perspectives are examined in this chapter. The analysis deployed in this study, however, favours a more pragmatic approach to modernisation. Such a pragmatic approach should not replicate Daniel Lerner's (1964) 'revival reformist approach', which is based on the Western model, but an independent approach to modernisation that takes cognisance of indigenous cultures as building blocks for development. Although several variables including, ecology, politics, and economic policies influence development, culture provides the logical explanation through which the interplay of these variables can be properly understood.

Development is a complex and multifaceted process that encompasses both individual and group aspirations (Rodney, 1972). In view of this, a soft definition will be the most appropriate analytical tool. Development has been defined 'as a societal shift from state of being to a desired state' (Adesina, 1994, p. 183). This desired state, associated with improvements in living standards, are multidimensional and tied to certain structures. The structural changes, according to Himmelstrand (1994, p. 18) are linked to six criteria: a society's capability of using its natural and human resources to feed itself even under the threat of mounting population pressures; the given society or country should be in a position of producing or making available the basic tools needed for food production; there should be, through popular consent, binding shared rules regulating the relationships of human activities in the particular system; there must be indigenous entrepreneurs and labour capable of propelling development, there must be an appreciable balance of trade, without extreme 'unequal exchange,' finally, a considerable amount of autonomy is necessary in the political and economic fronts, limiting the dependence on international monetary donor agencies, who often tend to dictate a country's internal events.

A cursory look at the performance of African countries shows that there is hardly any country in sub-Saharan Africa that has met the six conditions. In Nigeria, only two of the six prescribed conditions appear to have been partially met. With an estimated population of over 110 million[233] people, a multiplicity of ethnic groups, geographically diverse vegetation and crude oil production of 2.2 million barrels per day; the petroleum industry in Nigeria has a potential of lubricating a viable economy. Nigeria appears to have enough human and material resources capable to at least feed itself and provide meaningful development. Instead, poor planning and mismanagement of public resources have combined to make the realisation of these noble goals impossible, at least for now.

Second, although the Nigerian 'population has a superabundance of the entrepreneurial attributes',[234] it seems that their productive efforts have constantly been retarded by myriads of internal and external constraints. The other four criteria, which require a basic level of technological advancement; popular consent to shared rules; favourable terms of trade; and democracy presently are elusive and difficult to achieve. Nigeria, in contrast, can be described as economically underdeveloped, to use more mundane terms. The challenges therefore lie in reversing current trends.

Dependency and World System's Perspectives

Dependency and World System perspectives on development in non-industrialised countries are Marxist or neo-Marxist in outlook. According to either of these perspectives, the external dominance or capitalist penetration is the starting point of Africa's underdevelopment dilemma. Dependency theories emerged as a critique of modernisation theories. From dependency perspectives, the development process of less developed countries cannot be studied in isolation from the development of the advanced capitalist system. The economic predicaments of Nigeria and other less developed countries (LDCs) are viewed within the context of global economic relationships, in which the latter have been marginalised. For Frank (1969) 'economic development and underdevelopment are the opposite faces of the same coin. Both are the necessary result and contemporary manifestation of internal contradictions of the world capitalist system'. Following from this, expansions of capitalist systems or core zones are considered as inimical to the progress and development of the peripheries.

Dependency theories reject the notion that underdevelopment can be remedied through growth or modernisation promoted by capitalism and foreign aid (Frank, 1969; and Rodney, 1971). The contradictions of the social relations and the distortions imposed by imperialism are considered as important in analysing developmental problems. In this context, economic development and underdevelopment are seen as having 'a clear historical reference' (Phillips, 1977, p. 10). Historically, the devastating effects of the trans-Atlantic slave trade on the entire continent of Africa are too well known to warrant lengthy and superfluous discussion. Nevertheless, 'the removal of some twenty million people from any space is not a mere quantitative horror, it is an act which empties a living space in history' (Soyinka, 1992, p. 206). Significantly, most of the people lost to this iniquitous human traffic were the most capable men and women, among whom were productive farmers, skilled artisans, and industrious entrepreneurs.

Second, after the abolition of slavery, the African continent became a political laboratory for testing ideologies. Most of these experiments had devastat-

Table 11.1. Africa's External Debt and Debt Related Statistics (Billion of US$ and percentage)

	1993	1994	1995	1996	1997
Total debt (US $ Billions)	301.7	212.2	329	340.6	349
As a percentage of GDP	65.4	66.3	68	67.8	67.5
As a percentage of XGS	345.6	302.1	304.9	293.4	283.9
Debt Service (US $ Billions)	37.7	38.3	32.9	31	33
As a percentage of XGS	28.3	25.8	30.5	29.3	21.3

Source: World Bank 1998

ing effects on newly independent nations. During the colonial era, not only were political decisions left in the hands of colonialists, but economic decisions as well. Third, the mechanisms helping the modes of unequal exchanges are still in place even in the post-colonial period, permitting the exploitation of the developing world by the core zones (Chase-Dunn, 1981). Technological advancements in the core zones have led to the development of artificial and synthetic raw materials, which coupled with the high prices of capital goods results in very low prices of cash crop commodities, by that worsening further the terms of trade (Chambua, 1994, pp. 42–43).

The area of debt servicing provides a quantitative measure of Africa's economic problems. Africa's daily debt servicing to the industrialised world amounts to four hundred million dollars.[235] The dependency and world system perspectives perceive the debt factor to be one reason for the inability of most developing countries to disentangle themselves from their present economic problems. The consensus among development scholars is that, Africa's debt overhang poses a major obstacle to economic recovery and sustainable growth.

Table 11.1 shows that between 1996 and 1997, the external debt of Africa rose from three hundred and forty billion US dollars ($340 bn), to three hundred and forty-nine billion dollars ($349 bn), representing an increase of about three percent in one year. During the same period, Africa's debt service increased from thirty-one billion US dollars ($31 bn), to thirty-three billion US dollars ($33 bn), absorbing 21.3 percent of all export earnings. Debt alleviation is currently a major burden on the agenda of African policy makers. Heavy debt burden not only poses a threat to economic growth, it forms a major factor that increasingly raises the level of uncertainty and continues to deter foreign investors from investing on the continent.

In Nigeria, the situation is not different. In spite of the disagreements between the Nigerian government and the World Bank officials over the exact size of Nigeria's external debt, there is however, a consensus on the fact that the amount is colossal. Nigeria's total external debt stock based on an amalgam of debtor and creditor data by the International Monetary Fund (IMF) in August, 1997, stood at $28.7 billion.[236] External debt service ratio during the same year amounted to 13 percent, which was equivalent to Nbn 41.3.[237] Heavy debt servicing for many years has had a negative influence on real Gross Domestic Product (GDP).

It is significant to mention that all the subregions in Africa recorded lower growth rate in 1997 when compared with the previous year 1996. The lowest growth was however, recorded in Southern Africa, 2.4 percent and North Africa 2.8 percent. The highest growth rate of 3.8 percent however, was recorded in central Africa. In West Africa, the decline in Gross Domestic Product (GDP) was attributed to the economic recession in the larger economies of the subregion such as Nigeria, Ghana, and Cote d'Ivoire. In Nigeria, the decline in international oil prices was a major factor whereas in Cote d'Ivoire and Ghana the decline in agriculture, which is attributed to the impact of global *El-Nino* climatic conditions.[238]

The new dependency thinking, however, perceives the activities of foreign investors and market imperfections to be central to the economic stagnation of developed countries. In this relationship, the role of multinational firm in contributing to the net outflow of capital in developing countries is strongly emphasised. In contrast to this argument, neo-conventional writers see multinational firms as beneficial through, job-creation and as 'more responsive to legal and tax requirements than their domestic counterparts' (Biersteker, 1978, p. 51).

Nevertheless, the economic successes and the growth rates of peripheral economies such as South Korea, Taiwan, Singapore, Indonesia, and Malaysia have weakened the basic clichés of dependency arguments. The immense economic strides made by these countries in recent years have led Nigel Harris (1990), to predict '*The End of the Third World*'. Second, dependency and world system theories tend to perceive the world in binary terms, which is very often misleading. Moreover, an inherent limitation of dependency theory is its presentation of the African situation as helpless and unchangeable.

Fourth, the assertion that Africa's developmental problems are externally imposed absolves developing nations of responsibility for their internal contradictions. A recent survey of 4,000 entrepreneurs by the World Bank, out of which 3,431 respondents came from developing countries including sub-Saharan Africa, identified obstacles to development as largely local. The major local obstacles to development identified by all groups include corruption or lack of transparency in business, poor or high tax regulations, inadequacy of infrastructural development, and inflation.[239]

Internal obstacles and policy contradictions sometimes make developing countries high-risk environments for investment. In northern Nigeria, the relocation of peasants from ancestral homes to new locations on the Rima and Hadejia-Jamare river basins for government-funded irrigation projects engendered peasant resistance because of poor planning. The aftermath of the relocation of peasants in northern Nigeria for development purposes encountered deleterious consequences because policy planners ignored the aspirations of the indigenous people.

In Tanzania, internal contradictions were largely responsible for the failure of Nyerere's *Ujamaa programme*, a development programme that advocated self-reliance among its citizens while its architect became one of Africa's greatest single recipients of foreign aid (Nyang, 1992, p. 440). In both the Nigerian and Tanzanian examples, development efforts were stymied because of the government's failure to consider cultural and ethnic diversities. Internal policy contradictions have been responsible for the failure of public corporations in economic spheres where private individuals have performed creditably well.

A general weakness of dependency arguments is the common pitfall of ignoring the internal dynamics of Third World countries and the pervasiveness of the influence of the military. The military have been in the forefront of shaping policies since independence in most African countries. In Nigeria, the military have been in power in all but ten years of self-rule. The unprecedented abysses of military coups and annulment of the electioneering process over the years have proven counterproductive and incompatible with development efforts.

Prolonged military rule not only weakened the foundation of the democratic process, it also deepened inter-ethnic suspicion and mistrust in the Nigerian federation. There is now also a consensus among scholars that economic development cannot be divorced from political stability. Consequently, long-term political stability has been a strong factor behind the economic miracles of Southeast Asian countries such as Malaysia, Indonesia, and Hong Kong. The absence of political stability and an enduring legacy of democracy in Nigeria imply that political risk factors continue to affect cultural values and economic performance negatively. For an economy to do well, there must be stability as a *sine qua non* conditions.[240]

To sum up this section, it is justifiable to conclude that dependency theories often gloss over the crucial question of Africa's internal contradictions. In particular, the socioeconomic ideologies which dependency theories tend to propagate, such as socialism, labour movements, class solidarity, peasant movements, labour unions, and social activism, are weak in Africa (Mazrui, 1992, p. 133). While dependency theories are strong in pointing out Africa's problems, they are rather weak in propounding convincing solutions. Consequently, a considerable amount of development literature from dependency perspectives is often focused on the utopian questions of desirability (Phillips, 1977, p. 7). In view of this, a reconstruction of the modernisation theory from a cultural perspective is crucial.

Modernisation Perspectives

The modernisation, or orthodox development theory, addresses the question of how to combat underdevelopment and move towards a level resembling that of the advanced industrial countries (ibid.). The theory perceives all developing societies as undergoing transformations from traditional economies to the level of advanced capitalist countries (ACCs). It conceives the ACCs as having passed through the experiences of Africa and the larger part of the developing world. The theory proposes that Africa must undergo stages of development similar to those that Europe and North America have passed through (Chambua, 1994, p. 38). Stages that will lead to the integration into the world economic and social order include, population adjustments, literacy, with an emphasis on science and technology, new investments and trade patterns. Development is therefore taken as bridging the gap in these key areas. The quest to bridge the gap by most African countries during their early years of independence led to blind imitations of the industrialised world. For Chambua (1994, p. 38), a major limitation of modernisation theories is their tendency to disregard transnational structures that affect the changes in LDCs.

The assumption by early modernisation theorists that all countries follow the same path from traditional to modern, without regard to the peculiar internal structures of specific countries, contributed to the failure of development projects in sub-Saharan Africa. The path of development is viewed from a linear increase in national income with attendant capital accumulation, industrialisation, urban expansion, technological progress, and integration within the world market. Most development projects fail to yield dividends because they are premised on the notion that African societies are 'cultureless'. In view of this, the cultural settings into which new technologies are introduced are frequently ignored.

Yet, cultural understanding about societal attitudes towards obedience to authority, governance, democracy, and government property are relevant to development. In development-related discourses, a society's attitudes towards sustaining both the new and the old should be well understood as a tool for building new projects.[241] There are currently gaps in this area that often manifests in recklessness toward government property and developmental projects.

Culture is relevant to development because it has a utility function regarding the choice of the means of production of goods and services. Development discourses are generally focused on macro issues. A link with culture bridges the gap by allowing researchers and programme implementers the capacity to bring their individual cultural experiences to bear.[242]

Through cultural understanding of modernisation, societal sexual division of labour, traditional soil conservation practices, risk related attitudes, entrepreneurial ethics, trust and performance in modern manufacturing ventures can be properly understood. The utility of macroeconomic policies lies in their application within a specific cultural context. The manner in which both workers and their employers in the plastic manufacturing sector confer meanings, employ local techniques and adapt to modern manufacturing methods are embedded in culture. Such an understanding can be enhanced by integrating cultural analyses into the modernisation theory. Through cultural understanding of modernisation, policy makers will be in better position to know local values and priorities capable of enhancing or retarding a policy programme in a given society. The relevance of cultural study lies in its application as a diagnostic tool. Thus, for Wildavsky (1992, p. 153), questions, such as what is the problem in a country, what went wrong with a particular development effort, and what are the resisting agents can be understood only from a cultural perspective (ibid.). Good development projects often fail in the developing world because they are often foisted on recipient societies without recognition to cultural peculiarities. However, through cultural understanding of modernisation, development planners in Nigeria and other parts of Africa, probably can know in which periods of the year the application of a development method may be easier or more cumbersome.

The modernisation theory from cultural vantage points, which this study favours, perceives genuine development as progress generated in large part from below. Thus, for Ake (1988, p. 12), the failure of dependent development has made building on indigenous structures a necessary alternative for self-reliant development. Development must accommodate itself to the values and norms, interests, and social institutions of a society. Sustainable development implies that societal change has come to terms with culture (ibid.). Despite the relevance of culture in enhancing development, modernisation theories have glossed over its utility function over years.

A common approach in modernisation analysis is the construction of development programmes from a macro perspective. Development is viewed as an externally imposed, 'top-down' structural change. The dependent approach to development has often appeared too elitist and therefore frequently shunned by many Nigerians for which it is meant to assist. The basic flaw of most development projects has been their overemphasis of the public sector at the expense of the private sector, which is typically the more vibrant economic sector in developing countries. Modernisation theories of the contemporary era are not entirely deficient. However, their neglect of the internal dynamics of specific countries is one of their major handicaps. Some internal dynamics that negate development effort especially in Nigeria include, among other things, corruption and business fraud.

Corruption and Business Fraud

With the current institutional transformations in parts of the industrialised world, the role of corruption in economic underdevelopment is getting old-fashioned in these areas. Such assumptions, however, cannot be made of developing countries where fragile political and economic institutions have contributed to resource mismanagement, capital flight and state bankruptcy. Aboyade (1971) cautions that 'to pretend corruption does not exist only worsens the problem of resource allocation' in the developing world. For Adigun (1992), any analysis on culture, and especially, its role in economic development, would be incomplete without some discussion of the role of corruption.

In this section, an attempt is made to develop an analytical explanation of corruption and business fraud in the Nigerian business sector within an institutional framework. Institutional rules shape and govern human interaction in the social, economic, and political spheres of every society (North, 1996, p. 3). In addition, they shape societal evolution by helping to reduce uncertainty through the provision of structures for everyday life and the enhancement of performance (ibid.). Institutions, which are both formal and informal, differ from one society to another. These peculiarities account for the differences in social, economic, and political transactions between countries like Nigeria, the United States of America, and Japan. Formal institutions are those devised by human beings such as codes of conduct, constitutions, judiciaries, the police, military and various arms of government. Informal institutions, on the other hand, refer to the conventions and codes of behaviour that often arise without forethought or collective agreement (Hechter, 1987).

The intrusion of informal constraints, such as ethnic and social factors, on policy matters have led to inconsistencies in the application of formal rules and sanctions. The result of these institutional inconsistencies has been public distrust and noncompliance with formal rules. Advanced industrial countries differ in many respects from developing nations. A comparison between them often obscures rather than clarifies matters. However, such comparison only makes sense within an institutional context. Conversely, despite the sustenance of political corruption on a massive scale, the prevalence of efficient formal institutions and an entrenched culture of trust have contributed to the economic rise of the United States of America and Japan.[243] The formal institutional checks and balances in these countries refer to various arms of government and reliable judiciaries capable of enforcing sanctions against noncomplying members of the society no matter their status and social background. Developmental policies require enabling political, economic, and social environments to perform. In their absence, development suffers under institutional constraints.

Conceptual Problems

What the average citizen regards as corruption or fraud varies from society to society, culture to culture, and even class to class (Waterbury, 1976). This therefore implies that, in the discourse of corruption, no absolutes can be advanced. These conceptual problems often arise because there are potentially hundreds of acts that can qualify as corruption (Odekunle, 1982, p. 33). These include the introduction of fictitious or 'ghost names' in salary vouchers, false account claims by employees in firms, and kickbacks. The varieties of acts that are used interchangeably with corruption entail several slippery methodological and theoretical problems (ibid.).

Corruption can be defined as 'the manipulation of coercive potential of one's social position to gain personal or group advantage at another person's expense' (Anikpo, 1982, pp. 59–65). This is similar to business fraud, which refers to deliberate deception or trickery employed in business transaction. By 'its internal logic, corruption violates' the normative values associated with fairness, justice, liberty, or equality, which is an offence to 'society's moral sensibilities' (ibid.). In every act of corruption, someone always gains at the expense of another. Individuals, groups and entire nations may suffer from the consequences of corruption. This is the reason corruption is frowned upon by all cultural groups in the world.[244]

Van-Roy (1970, p. 87), observes that, despite its universal abhorrence, corruption is a universal phenomenon. For Leys (1965, pp. 215–230) 'there seems to be widespread feeling that the facts cannot be discovered, or if they can, they cannot be proved, or if they can be proved, the proof cannot be published'. Corruption in this text is treated as one of the major obstacles that hinders smooth transaction in the Nigerian business sector.

Although a variety of acts may qualify as corruption, the nature, purpose, and frequency with which such acts are committed depend on the cost of corruption in the particular system. The higher the cost, in terms of penalties for offenders, the less corruption there will be in a given society. That corruption has become pervasive in the Nigerian socioeconomic and political system is the result of an absence of costs in terms of punishment or embarrassment for key offenders. Contemporary decadence in formal institutions and the decline in public accountability in Nigeria furnish the feeling that societal rules can be broken with little or without cost.

A recent survey of 54 nations by the *Washington Post* ranked Nigeria as the most corrupt nation in the world. Among the least corrupt nations are New Zealand, Denmark, and Sweden.[245] Nigerian con men are alleged to have duped many foreign business people of colossal sums of money in shady deals involving nonexistent contracts. Although some of these claims and negative labels may have been over-exaggerated, they underscore the extent to which corruption has penetrated the political and socioeconomic fabric of the Nigerian life. The absence of public accountability makes Nigerian contracts more expensive and 200 percent higher than in Kenya.[246]

Several reasons have been advanced about why people engage in corruption and business fraud. According to Akinkoye (1982), scarcity creates the necessity for using illegal methods to achieve ends. In a system with overwhelming unemployment problems, the scarcity of jobs could create the urge for applicants faced with stiff competition to use illegal means in order to be employed. Corser (1967) and Mensah (1982) have observed that the denial of basic legitimate rights to the citizenry triggers the use of illegitimate channels to achieve ends or to make up for losses.

The pervasiveness of corruption in the Nigerian society has given birth to a variety of local terms. In recent times however, the most common term for describing corruption especially in Nigerian business circles is Advanced Fee Fraud (AFF), synonymous with a criminal code known as '419'.[247] Other local equivalents include, *long-leg*, *chop-chop*, *bread-power*, *awuf*, *emergency contractor*, *arrangee*, and *ten percenter*. Among the Hausa, varieties of linguistic terms synonymous with corruption include *cin hanci*, *rashawa*, and *kashe mu raba*. Smith (1964) argues that the existence of these linguistic terms before the encounter with the colonial world is evidence that corruption is not foreign, but endemic in Hausa culture. Accordingly, corruption is a product of lax supervision and the absence of accountability in a specific society (ibid.).

Smith's (1976) study however, fails to acknowledge the fact that, the legacy of forced labour, monopoly of cash crop exports, and the illicit self-enrichments of colonialists, became the integral part of native authority and political machinery after independence. Corruption in the Nigerian public sector is a hangover of the British colonial heritage. Simpkins and Wraith (1963) have argued that corruption has been with societies throughout history. In reality, corruption is universal to all societies, including the industrialised ones. However, its nature, intensity, pervasiveness, and control may vary from one society to another. Thus, corruption is considered a stage in development. Corruption gains ground as a society enters the critical phases of modernisation (Clinard and Abbot, 1973, p. 200). Thus, modernisation opens new opportunities for local contracts, which fuels corruption and fraud among local technocrats and contractors. There appears to be a link between corruption and a society's level of development.

Societal perceptions of corruption in Hausa society, as elsewhere in Nigeria, are mixed and rather unclear. The extent to which corruption is considered evil or negative depends on the context or situation in which it occurs. Whereas honesty is generally regarded as virtue, it may sometimes be seen as an encumbrance. Writing on societal perception of bribery, Ekeh observes that:

> A man who holds office in the civic public and upholds the norms of impartiality in the exercise of the powers and privileges of his position by being evenhanded towards all groups is bound to be regarded as disloyal to his primordial group. On the other hand, an official who does special and favoured benefits for the interests of his primordial group is hailed as a hero by his people even if he is proven guilty and condemned by the courts of the civic public (Ekeh, 1994, p. 245).

Corruption then is often a product of the confusion between societal modes of adaptation to goals and the means used to attain them. Where the means and especially the goals are ambiguous and unclear, various forms of deviations set in.[248] The two polarised dimensions of corruption are, its use for personal gain and its manipulation for societal and primordial interests. In Nigeria, people in privileged positions are constantly under pressure to 'enrich their primordial public' and kin groups 'from the coffers of the civic public' or from the business firms in the private sector where they work (ibid.). To understand the negative or positive role of corruption to development, one must understand the context within which corruption takes place and try to qualify explanatory variables.

The functionalist approach considers corruption as having a utilitarian aspect in its ability to soften bureaucratic rigidity in new states (Werlin, 1972, pp. 251–2).[249] In a developing country like Nigeria, it is rather a teleological argument about whether corruption increases bureaucratic rigidity or vice versa. Nye's (1968, pp. 106–118) study portrays corruption as a potential avenue for capital formation, a supply of incentive for local entrepreneurs, and a medium through which the elite can be integrated. In many industrially advanced countries such as Japan, the United States of America, Germany, Italy, and South Korea, corruption has become the medium through which multinational firms penetrate hostile markets or gain big government contracts at home. Corruption in the Nigerian society reveals a contrary picture. Recent government investigations have shown that individual gains in corrupt practices are seldom reinvested in viable economic ventures within Nigeria.[250] The use of foreign banks to deposit huge sums of hard currency negatively encourage capital flight and inhibits liquidity circulation necessary for stimulation of the local economy.

Studies by Myrdal (1968) and Weidenheimer (1970) have shown that corruption has no positive or utilitarian function. From both cultural and moralist points of view, corruption has no positive function, since it violates or deviates from societal rules for the sake of personal gain (Weidenheimer, 1970). Myrdal (1968) observes that 'corruption introduces elements of irrationality in plan fulfillment by influencing the actual course of development in a way that is contrary to the plan'. By virtue of its ability to change the course of planned action, corruption often leads to hurriedly taken measures and the production of poor and substandard goods and services in a society.

Institutional Context of Corruption

By institutions, I refer to rules devised by human beings for the smooth transaction in society. Institutions therefore cannot be seen, touched or measured, 'they are constructs of the human mind' (North, 1996, p. 107). Institutions provide the formal mechanisms on which to build trust in a society (Sanner, 1997, p. 130). For North (1996, p. 1–10), the poverty in third world countries is the result of institutional constraints which define a set of payoffs to political/economic activity that does not encourage productive activity. Stability of the banking and financing sectors is crucial in this regard. In sharp contrast, their failure could lead to a destabilisation of local currencies and economic activities. The reliability of insurance companies and fulfillment of financial obligations are necessary in providing the safety of economic property against the unforeseen contingencies. That there is low level of trust in the Nigerian society does not imply that Hausa or Nigerians are untrustworthy. However, the institutional framework as currently constituted does not positively encourage or allows network on political capacity to capture the potential economic gains.

In Nigeria, the public sector has frequently defined and provided the basic framework for controlling corruption in all sectors of the economy. Nearly all previous administrations, including the current one under General Abdul Salami Abubakar, have introduced formal institutional codes, public enlightenment campaigns, and laws regulating corruption in their social, economic, and political agendas.[251] That corruption continues to expand on a wider scale, is not a product of lack of effort, but a reflection of insincerity by the Nigerian leadership. The expansion of corruption among the inner circle of government reveals the internal contradictions of the state's anti-corruption measures.

It is a big paradox that, the state structure in the forefront of anti-corruption campaigns in Nigeria is itself riddled with corruption and nepotism. As such, every new administration in Nigeria owes the public a duty to probe its predecessor. The recent probe of high ranking government officials and the recovery of $1 billion US dollars and over thirty-five houses from former aides of the late General Sani Abacha is a testimony to the pervasive nature of corruption in the country.

Deterioration of the Nigerian civil service has amounted to a personalisation of public property and pocketing of state funds by key government figures. This ugly scenario continues to rob society of crucial developmental projects. A report by the Federal Ministry of Finance confirmed illegal withdrawal of N1,431 billion through the Central Bank of Nigeria (CBN) a former national security advisor (NSA).[252] Today, public mismanagement in the oil sector and illegal lifting of crude oil to overseas countries by high-ranking government officials has led to acute local shortages, prohibitive increases in transport fares, and abysmal rises in inflation.

Like in many African countries, the evidence in Nigeria suggests that populists' anti-corruption measures often serve as a justification for military intervention or interruption of the democratic process. Under such circumstances, the targeted persons for government's anti-corruption measures are old political foes. Conversely, highly corrupt ethnic members and political allies are frequently bypassed by formal judicial and criminal sanctions. Policy inconsistencies continue to fuel ethnic and religious suspicion in the civic society. The result of societal distrust in official policies and sanctions has led to non-compliance and virulent corruption in both public and private spheres. This seems to suggest that the anti-corruption measures have failed.

Economic institutions need a politically enabling environment to produce desirable results. The way business operators relate to the state is also important. In this regard, the role of government in awarding contracts, allocating foreign exchange quotas, and formulating labour laws, cannot be ignored. The dwindling government revenues accruing from erratic changes in world oil prices resulted in a scarcity of foreign exchange. This scarcity has increased bureaucratic rigidity and corruption in foreign exchange allocations to the business community.

Evidence in the three plastic firms suggests that rather than relying on the state for foreign exchange allocations, the business proprietors of the plastic firms got more than 70 percent of their foreign currencies through direct purchasing from private money bureaus. As in many African states, access to friends, ethnic and religious members in key government positions is crucial to business transactions in Nigeria, although this has not been the case with the three firms covered by this study.

Prolonged delays often experienced in obtaining foreign exchange allocations imply that sometimes manufacturers have to wait for several months to be able to import raw materials from overseas countries. Usually, the time lags between the securing of foreign exchange from banks and procuring raw materials abroad for production and sales are very long and unsuitable for manufacturing activity. The problems relating to foreign exchange allocation are further compounded by the official deregulation of banks and the valorisation of the naira. These developments have resulted in two tangentially opposed markets in which monetary transactions and foreign exchange deals are conducted. These are the autonomous foreign exchange market (AFEM) and the unofficial, or illegal, markets. Of the two markets, the latter is more vibrant. While the government exchange rate is fixed, it is flexible at the commercial banks and unofficial currency markets. Sometimes the rate is quadrupled in the unofficial market although such institutions were supposed to add only a one-percent markup.

This development had several implications for the banking sector. The number of banks in the country during the last decade has increased from 40 to 116. Money laundering, cheque-kiting, and other fraudulent practices have become common in the banking sector. Consequently, most commercial banks in Nigeria are more eager to deal with men who transact in the illegal foreign exchange market than with genuine business people. Artificial scarcities are therefore often created in the banks to compel buyers of foreign exchange to go to the illegal market where the banks have agents. The reselling of foreign exchange allocations has become far more profitable than any productive manufacturing activity in Nigeria. Cheap profits accruing from the sale of foreign exchange allocations by business industrialists in the financial market continue to stifle productive manufacturing ventures.[253] Despite the prevalence of anti-corruption measures, the contradictions in state political, legal, and economic policies have encouraged the expansion of corruption on a wider scale.

Implications for Trust in the Business Sector

Honesty and reliability are necessary components for progress, in the economic sphere. Conversely, improbity and lack of integrity are associated with stagnation and bankruptcy. For many years, the Nigerian business sector has been 'derided as an investor's nightmare'[254] because of corruption. Frequent macro policy changes and absence of endurance of specific patterns or relations may seem to negate the construction of trust. The negative effects of corruption and derailing of macro policies on the Nigerian business sector are colossal and varied. First, the weariness about the safety of imported goods and the cost of bribing government officials at the nation's ports continues to scare away both foreign and indigenous investors from the Nigerian business scene. Second, expenditures on bribery are sometimes passed to consumers in higher prices of goods and services.

Frequent contravention of liquidity rules by banking officials in Nigeria has led to huge financial losses, and crises, in the banking sector. Corruption and poor performance of banks compelled the Federal Government to create a Failed Banks Tribunal in 1996. In the first quarter of 1996, forty-two Nigerian banks were classified as distressed. In view of this, 17 banks were put under the direct control of the Central Bank. There are currently 40 directors and managers of Nigerian banks in jails awaiting trial on allegations of corruption and financial misappropriation charges.[255]

The evidence in Kano and other parts of Nigeria suggests that one reason for the aversion of capital loans from modern banking institutions by indigenous entrepreneurs is the result of the difficult corrupt practices involved in securing them. For every successful loan that is obtained, the beneficiary is obliged to part with a certain percentage as token gift, or *goro,*[256] to occupants of key positions in the banks. The 10 percent fee or gift depends on the size of loan and the banking institution involved. This implies that, in actual practice, there is always an additional cost to a bank loan. The high costs of local goods sometimes make imported items cheaper and more attractive than locally manufactured products. This trend has frequently stifled the growth of indigenous manufacturing businesses.

Corrupt practices in the Nigerian business scene also account for the circulation of huge amounts of counterfeit and bogus currencies in the economy. The difficulty of controlling the circulation of counterfeit money, or *kudin jabu*, in Kano and other parts of Nigeria stems from the fact that victims of spurious currency transactions usually pass their misfortunes and cost to unsuspecting customers, rather than report to unreliable government officials. Excessive circulation of spurious or bogus cash in the Nigerian economy continues to generate a spiralling wave of inflation, which frequently renders government macroeconomic policies and development plans meaningless and ineffective.

The pervasiveness of corruption in the private commercial sector has been responsible for the sale of fake, expired and adulterated foods which have resulted in innumerable public health hazards. Consequently, local consumers are often fearful and suspicious of locally manufactured products. This has resulted in the low patronage of homemade products where foreign alternatives are available. Sometimes the intermediary role of unsuspecting local merchants in the sale of substandard and adulterated goods puts them in very awkward and embarrassing positions.

In the organised private sector, the preponderance of business fraud and financial scandals has been responsible for the closure of several micro-businesses in Kano. Furthermore, embezzlements of corporate funds by accounting personnel and the disappearance of sales people with sales proceeds and company vehicles to unknown destinations are common phenomena in most Nigerian firms. In recent years, photographs of wanted fraudulent employees feature regularly in Nigerian newspapers. The evidence presented in the fifth chapter of this study shows that both the Lebanese and Hausa employers of the plastic factories were prone to nepotism in their recruitment of employees through ethnic and social networks. This practice resulted in the recruitment of often incompetent or unproductive workers. Informal sources from the three plastic factories suggested that there were rampant cases of bribery between managers, custom and union officials.

Approximately 54 percent of the employees confirmed that they had experienced various forms of fraud or corruption in their places of work whereas 46 percent of them showed that they had never experienced any corrupt practices in the firms. Interestingly, two-thirds of the employees who never experienced corrupt practices were supervisors. This perhaps is a reflection of the sympathy that most supervisory and intermediate staff in Nigerian factories have for management. Significantly, a large degree of corrupt practices, representing 30 percent, was experienced by the junior employees.[257] In the factories acts of corruption include bribing supervisors to influence managerial recruitment of kin and townsmen and for covering absences, lateness or major mistakes committed in the production process. Most acts of corruption experienced by junior workers were often minor ones which did not include the major bribery deals that are concealed or hidden by managers and business proprietors.

In the Chinese factory, 13 percent of the employees observed managerial arbitrary allocation of plastic products to key union leaders. Such gestures were occasionally made by expatriate employers in the plastic factories to win the support of key union members over internal matters. Most of the illicit financial remittances between business proprietors and customs officials usually ended in personal pockets rather than in government coffers. In the Chinese factory, arbitrary rewards by business proprietors to recalcitrant union members, and selected employees were very rampant. Arbitrary rewards only served to fuel discontent among junior employees rather than encourage productivity. There is no evidence in Kano or any part of Nigeria that the gains from corruption were reinvested in gainful economic activities. On this account, one can conclude that the effects of corruption on the Nigerian economy are wholly negative. The discussion in this Chapter suggests that the eradication of corruption does not rest on adequate provision of employment, or goods and services alone, but on building of efficient institutions, and long-term education of the citizenry about the ills of corruption in nation building. Above all, a transparent leadership as a point reference can be a good starting point for the building of a honest citizenry.

12 Summary and Conclusion

Entrepreneurs and Societal Progress

Research on the role of entrepreneurs in economic development provides two contrasting positions. The first perceives entrepreneurs as exploring uncharted economic territories, discovering new market niches, providing jobs and improving the general standard of living in their societies (Hayek, 1980a; Kirzner, 1983; Ikiara, 1994; and Woodward, *et al.* 1994). From a Marxist position, the entrepreneurial profit motive and the capitalist system in general would seem to be considered as greedy, exploitative, and non humanitarian in a manner antagonistic to societal as well as economic development.[258] The latter departs from the findings of this study. It also deviates from the large body of literature that underscores the relevance of entrepreneurs in societal development. In spite of the presumed selfishness of entrepreneurs, many if not most 'crave being part of larger communities'.[259] Ideal entrepreneurship in Hausa society, is the type that effectively balances economic considerations with religious motives without unnecessary costs.

Although Weber (1947) may be of a view that economic progress is possible in non-western societies, he is however, blunt in pointing out such progress less likely in an Islamic society. The reasons according to him are first, the presence of pervasive violence, which continuously threatens and obstructs both the economic progress and social order. Second is the erroneous view that the economic or merchant classes in Islamic societies were too weak to wrestle power from the authorities. Although I do not intend to reiterate a critique of Weber on the notion of violence in Islam, nevertheless this aspect of Weberian analysis has been over-exaggerated. Conversely, the records show that in the early years of Islam, successful and powerful Arabian merchant classes played crucial roles in the spread of the religion.[260] Thus, Weber not only underestimates the role of urban merchant classes in shaping the course of Islam, he also completely ignores the peculiar cultural characteristics of Muslim societies across the world.

In northern Nigeria, and especially Kano, entrepreneurs are the distinguished classes. It is these powerful economic classes who rub shoulders with the ruling class. The evidence in Kano and all over Nigeria suggests private entrepreneurs play significant roles towards societal progress. I agree with Tahir (1975) that a very large aspect of Weberian arguments concerning Islamic militarism and the dominance of military groups in the political economy do not apply in Kano.

This study has also found some variants of the so-called Protestant economic ethic in Kano and parts of Nigeria. The Protestant ethic is more widespread among the Ibos who are predominantly Catholic, than in other Christian religious groups. Furthermore, the economic progress of Muslim puritans of Kano is at best parallel with the Weberian thesis (1974). The evidence also suggests that Hausa and Islamic values support entrepreneurial resourcefulness. The religious economic ethic appears therefore very strong among the Muslim merchants of Kano.

In Weberian formulation, progress assumes an ideological character, which refers to a Western or European way of life.[261] Nevertheless, the evidence across the world shows that cultures differ in their realisation of progress or development (Montagu 1968). Put in the proper context, economic progress is possible anywhere. In Hausa society, progress may refer to incremental change. Progress refers to change that is in consonance with religious and cultural values. Thus, progress does not require total jettisoning or discarding of the past, but building on its positive aspects. This idea of progress is similar to Hausa definition of business success examined in Chapter Ten. Building further on this analysis, the main fetters upon which societal progress can be identified include, above all, the presence of indigenous entrepreneurs and labour capable of harnessing the local natural and human resources to societal advantage. For Ani (1994, p. 497) changes 'occur in greater numbers where they are encouraged, even mandated by a culture that lives for them and by them'.

The economic transformation in Kano is partly due to the mixing of cultures from disparate sources. For many years, Kano has served, almost, as a 'melting pot' of sub-Saharan African cultures. The tremendous acculturation processes make the merchants of Kano more flexible and adaptive in their economic pursuits. The geographically central position makes Kano a more suitable focal point for business activity.[262] From a peasant and cash crop economy, diversification into modern industrial activity by local business groups accounts for Kano's modern economic prominence. Although progress or development has universal and local referents, both will fit into the transformations that have taken place in northern Nigeria, since the turn of the twentieth century. The period marked the integration of the region into the capitalist world economy.

Thus, from a peasant economy, Kano and many parts of northern Nigeria have witnessed significant transformation to a semi-industrial economy.

The evidence in contemporary Kano suggests that economic prosperity among Muslim Hausa groups transgress sectarian boundaries. There are, invariably, local successful merchants and industrialists among the different Islamic sects. Kano is unique in the sense that the annual inflow of migrants into the city by far supercedes that of any other northern Nigerian city. The supply of cheap migrant labour has enabled merchants to appropriate surpluses. A survey of major businesses across the world shows that migrants possess a tremendous potential for creating new economic niches (Waldinger, *et al.* 1990). This finding is also applicable to Kano where two-thirds of the prosperous merchants are migrants from the rural parts of Hausaland.

It is worth reiterating here that the voluntary diversifications of indigenous entrepreneurs into modern industry show they are rational. An essential characteristic of a rational entrepreneur is the pursuit of profit. Hausa entrepreneurs of Kano and elsewhere in northern Nigeria display this quality in great abundance. Through the pursuit of business profit, many private entrepreneurs continue to play significant roles in this transformation process, by shaping the development of their communities. The role of private entrepreneurs in seeking new economic opportunities has for many years kept society in motion. The creation of new market niches, products, and services by private entrepreneurs invariably has a logical connection with societal transformation which cannot be ignored. As risk-bearers entrepreneurs make a developmental contribution when they succeed (Ray 1988, p. 4). Not surprisingly, the performance of indigenous entrepreneurs has become a measuring rod for societal progress. Thus, the progress of indigenous entrepreneurs inevitably implies the progress of Kano.

When entrepreneurs undertake risk, they indirectly make sacrifices, not only for themselves and their immediate families, but for their societies as well. Entrepreneurs are generally considered as cost-effective agents of economic development. This is due to their ability to 'extract the very last bit of marginal capacity out of whatever resources that exists within society'. Although other agents of change may accomplish the same ends, they may not be as cost-effective as entrepreneurs (ibid.). The relevance of entrepreneurs lies in their role as agents for the development and management skills.

Experience has shown that some indigenous merchant capitalists who diverted into modern industry were not solely motivated on economic grounds. Consequently, the pride of personal achievement and the desire to demonstrate patriotism are among the reasons. In a study of Hausa merchants in neighbouring Niger Republic, Grégoire (1992) observes that indigenous merchants who diverted to modern industry were the successful ones who tried to show their nationalistic attitudes by accepting the call to provide jobs.

Investment records in the Kano industrial sector show that expatriate investment has risen from a modest beginning of 10 manufacturing and processing firms in the early 1950s to a peak of 77 new projects in the 1970s. By contrast, indigenous investment in industry remained steady from eight projects in the 1950s and rose sharply to a peak of 58 in the 1970s.[263] Although data on industrial investment in the 1990s is not readily available, the figures show that both expatriate and indigenous investments in this area have declined continuously since the late 1980s. Nevertheless, the positive effects of both expatriate and indigenous investments in the Kano industrial sector during the past four decades have been considerable. In the three plastic firms, it was observed that only four business industrialists were providing employment to some 820 employees. Large indigenous family firms such as the Gaskiya Textiles, Dantata and Dangote groups employ over three thousand workers in their economic activities. It is in this regard that the role of entrepreneurs in economic development becomes obvious.

In spite of the current economic reverses that have affected most aspects of the Nigerian economy, the contributions of indigenous and expatriate entrepreneurs, in the provision of employment and improving the standard of living has been quite substantial. With only a few manufacturing firms in 1950, there are currently over 350 manufacturing and processing concerns in the city of Kano, providing employment to an estimated 260,000 employees.[264] This does not include petty manufacturing, processing, and trading activities in the informal sector where market competitiveness increasingly provides jobs to thousands of the unemployed. Through rational pursuit of their economic activities, the entrepreneurs of northern Nigeria have gradually transformed society from a peasant oriented cash crop economy to semi-industrial society.

Prosperity means incorporating parts of economic liberalism into local cultural system.[265] In keeping with these demands, the public sector, until now considered a government domain, has been opened by the Federal Government for private participation. The telecommunication, power, and oil sectors have been slated for privatisation and private participation by the present military government. There is now substantial local investment in the telecommunication sector with the appearance of many indigenous telephone companies. Economic liberalisation and the empowerment of the private sector are surest means of energising the Nigerian economy. Today society is becoming increasing rational through modernisation and globalisation processes. Nevertheless, these variables do not operate alone, they interact with cultural variables relating societal work attitudes, entrepreneurial investment patterns, risk-related attitudes, and trust. In this regard, the role of culture in influencing entrepreneurial behaviour becomes more crucial.

Cultural Foundations of Economic Behaviour

This study has sought to bring to the fore, the linkages among culture, entrepreneurial ethics, and trust. Culture and religion are relevant because they provide rules and beliefs for what individuals and groups do. The data in this study confirm that cultural and religious factors play a role in economic behaviour in many ways. In Kano, and other Islamic societies, the open economy is predominantly a male domain. In spite of this, the economic household activities of secluded women in Kano and parts of northern Nigeria are substantial and comparable to the industrial wage bill. The earnings from the economic activities of secluded Hausa women provide sustenance in many homes. The evidence also shows that in the Islamic society women could maintain religious and cultural rules, work behind the harem and still be productive to their society. This, as the study has shown, forms part of the economic rationality of the predominantly Muslim society in northern Nigeria.

Managerial attitudes and recruitment patterns relating to gender in the plastic factories were influenced by cultural and religious factors. Recruitment patterns, especially in the Hausa and Lebanese firms were favourably disposed toward male and against female employees. Consequently, while male employees constituted 92 percent, their female counterparts represented only 8 percent of the entire population of the labour force in the three factories. A cross-tabulated profile of the female workers further reveals that approximately 81 percent who professed to be Christians, hailed from the middle-belt and southern parts of the country. A relatively small fraction of the Muslim female workers, representing 19 percent, came from northern Nigeria.

Cultural, historical, and religious factors reflect an overwhelming preference for commercial activities among the entrepreneurs of Kano and other Islamic societies. Observation in the commercial sector of Kano further reveals that entrepreneurial methods towards the acquisition of venture capital have a cultural genesis. Nevertheless, venture capital for large-scale modern manufacturing activity often comes from the modern banking sector. Intra-brotherhood religious solidarity in the business sphere, and the advancements of goods to traditional brokers, or *dillalai*, before financial settlements are economic habits embedded in cultural and religious values.

Hausa cultural perception and stigmatisation of risk as the equivalent of vice or gambling specifically weakens entrepreneurial enthusiasm towards economically risky activities. Entrepreneurial safety strategy against risk is reflected in the operation of multiple and unrelated businesses especially in the commercial sector. Apart from bureaucratic and collateral problems, the low level of local patronage of banking institutions is partly rooted in religious and cultural stigmatisation of interest on loans and financial deposits.

In the business sphere, cultural and religious factors account for the proliferation of mosques in business premises, including those of the three plastic firms. In the Lebanese and Chinese plastic factories, managerial disagreements with employees over the intrusion of 'prayer time' on working hours provide support for the claim that cultural and religious factors have an impact on economic and industrial relations in the factories of Kano. Ostensibly, the Islamic factor cushioned the effects of employee religious militancy in the Hausa and Lebanese factories because their proprietors were Muslims.

In the Chinese factory, the scenario was in many ways different. Perhaps the owners not being Muslims or Christians gave legitimacy to a very tough stance towards the factory workers. Among the factory employees, the fact that the Chinese did not belong to any of the locally practised religions also gave legitimacy to their hostility towards them.

Managerial alteration of working hours during the month of *Ramadan* to suit employee religious practices and coping capabilities in the three plastic factories, and indeed all factories operating in Kano, was influenced by cultural and religious factors. Moreover, managerial rewards and loan advances to factory employees during major Islamic and Christian religious festivals also suggest that cultural and religious factors permeate modern economic relations in Kano and other developing societies.

In theory, cultural values and norms encourage honesty, sobriety, hard work, credit worthiness, etc. However, in practice there are imminent signs of their decay or malfunctioning. The real challenges to development effort in Kano and Nigeria as a whole lie in the identification of positive indigenous values and putting them into good practice. The polarisation of the modern and the traditional raise questions on how the two can be efficiently balanced. Nevertheless, the extent to which new foreign economic values can be combined or compromised, as this study has shown, will inevitably depend on their compatibility with indigenous Hausa cultural values.

In spite of this, the findings in this study show that the role of indigenous and expatriate entrepreneurs in the economic transformation of Kano has been considerable. That many Nigerian businesses have not disappeared under the persistent pressures of political maladministration and environmental constraints only shows that many businesses operate outside the formal economy. The businesses in the informal sector frequently show a strong potential and a regenerative capacity. The informal sector of Kano and elsewhere in Africa therefore represents an intermediate stage, which if skilfully handled could become an important component of development policy.[266] The inadequacy of official monitoring and integration of the informal sector into the modern formal economy suggests that both sectors are not functioning properly and they interact negatively with each other. Therefore, a large volume of potential revenue in the informal sector, which could be utilised for development purposes and job creation, is lost annually.

The nurturance and promotion of indigenous entrepreneurs by future governments in Nigeria could be achieved by diversifying the activities of these micro-enterprises into those that complement the modern sector. In this way, the informal sector can be expanded into small and medium-scale enterprises and industries and begin to operate according to a set of dynamics largely activated by effective demand from the modern sector. A large volume of microeconomic activities in the informal sector, if well harnessed and integrated into formal institutional policy, can lead to job creation, economic independence, local innovation, and integration into the capitalist world economy.

The research findings show that indigenous cultural values positively produce entrepreneurs but not necessarily entrepreneurs in the real modern sense. This implies that entrepreneurial ethics and trust relations in their current forms are unsatisfactory. In spite of the prevalence of traditional micro-credit schemes, entrepreneurial preference for joint and partnership ventures in Hausa society remain very low.

Islamic and Hausa values generally encourage the pursuit of knowledge for avoiding pitfalls in all human endeavours. Although there are no established links between economic success and educational achievements, 'the way in which knowledge develops influences the perceptions people have about the world around them and hence the way in which they rationalise, explain, and justify that world' (North, 1996, p. 76). The development of knowledge will inevitably alter entrepreneurial attitudes towards joint businesses. Traditional and precapitalist micro credit schemes such as *adashe* and *asusu* can be improved and developed with modern rules governing their transactions. There are currently no properly developed rules governing the operation of traditional micro-credit institutions in Nigeria.

Yet, Islamic and Hausa cultural values constrain neither hard work nor resourcefulness. An observer of Hausa, and indeed Nigerian, entrepreneurs and workers in agricultural, commercial, and modern manufacturing activities will notice that they spend often more hours under the heat of the sun and in poorer working conditions than their counterparts in the industrially advanced world.[267] Their low productivity in agriculture and the slow pace in industry are often institutional problems associated with the poor management of resources due to ignorance. Fukuyama (1996) observes that 'modern wealth is based on human capital, knowledge, technology, innovation, organisation, and a host of other factors related to the quality rather than the simple quantity of labour used to create it'.

The empirically strong cultural thesis[268] about the necessity of a harmonious correlation between employee and management norms and values and high levels of profit has not been vindicated in this study. The Chinese firm, which exhibited more cultural disharmony and incongruence was rather highly successful by local standards. However, this single case does not allow us to generalise about cultural incongruence in relation to organisational progress. The evidence shows that employee perceptions about a firm's profitability influenced their demand for better remuneration and improved working conditions from their employers. Employee feelings about high profits and inadequate wages were the basic reasons for their discontent in the Chinese and Lebanese factories.

The research findings show payment of incredibly low wages in three plastic firms. The average salary of a factory worker in the three plastic firms was N1,200 approximately 13 American dollars per month. Although the payment of good wages may not be the only condition for productivity, employers of labour cannot pay extremely low wages and expect high productivity from their employees. A study by Brown and Nolan (1988) supports the idea that good wages are prime reinforcers of employee productivity. Employers of labour in Nigeria and other parts of Africa have to come to terms with the fact that one way of improving productivity is the payment of realistic wages to their employees. The payment of wages in all the plastic factories was based on a day's job rather than on employee productivity. This practice is seen in many industrially advanced countries as old-fashioned. In order to ensure higher productivity, Nigerian and Hausa employers must begin to assess and pay their employees according to their productivity rather than their mere presence at the factory yard. Payment of low wages as shown in this study constrained trust relations between factory workers and management relations in the plastic factories.

Three factors are seen as contributing to the poor treatment of workers in the factories of Kano and northern Nigeria. First, there is the absence of an effective monitoring system by the various arms of government to ensure that employers complied with labour regulations. Second, there is availability of surplus labour in the labour market willing to accept incredibly low wages in the factories. Third, a large proportion of the factory workers are poorly educated and quite often incapable of knowing what constitutes employee rights in the modern factory. Some of these factors were responsible for the common labour abuses in the firms. Although prolonged employee exposure to plastic flakes is hazardous, none of the three factories provided safety equipment or protective gear to its shop-floor employees.

The study also found that apart from the Chinese firm, neither the Hausa nor Lebanese plastic firm was unionised. Hausa cultural emphasis on social harmony in working relations, as observed in this study, is generally antithetical to the heterogeneous and pluralist interests which unions tend to represent. Entrepreneurial cynicism and abhorrence of employee unions were often derived from a desire in paying low wages. The absence of union in the Hausa and Lebanese firm coupled with the presence of a very weak union in the Chinese firm contributed significantly to the low degree of trust among workers and management in these factories. Prolonged military rule in Nigeria and the absence of a democratic framework with relevant ideology have further weakened employee unions in the firms. Before this, the activities of the populists Northern Elements Progressive Union (NEPU) and People's Redemption Party (PRP) provided the fulcrum on which employee unions revolved. Without a relevant ideology, most local employees in all the plastic factories relied on spontaneous or inchoate ethnic and religious sentiments as a means of advancing their grievances.

This comparative study of Hausa, Lebanese and Chinese firms, has suggested several questions about Hausa, Lebanese, and Chinese management styles in Kano which need further research enquiry. The research question is: under what conditions do family enterprises exhibit the favourable traits found by Church (1993) or turn out meagre results as shown in the study by Kets-de Vries (1992)? This controversy over family firms which was raised in the sixth chapter of this study, needs further explication, especially in contemporary Kano where most of the businesses are operated by families. The prospects about whether there are more issues at stake about entrepreneurial distrust for modern joint or partnership ventures than have been dealt with in this study deserve further research investigation.

The current trends in indigenous craft activities, their changing patterns, and the prospects to which they can be transformed to complement the modern industrial sector also deserve greater research attention. In particular, the complementary role of local garbage collectors to the plastic industries in Nigeria is worthy of investigations. There is currently no study of this nature. This study has only done that in passing.

The study also suggests that a modernisation theory from cultural vantage points will address genuine development in large part from below. This implies building on indigenous structures as a necessary alternative for self-reliant development. Sustainable development has to accommodate itself to the values, norms, interests, and social institutions of a society. It also implies that societal change has come to terms with indigenous culture.

Thus, Culture can reinforce development in many areas. First, through public enlightenment, family solidarity can be extended to embrace both enterprises and the nation. In addition, cultural and religious tenets can be used to encourage or enforce contractual bonds or credit worthiness.[269] In this way, traditional methods of capital accumulation and credit rotating schemes can be regulated with formal established rules and expanded into indigenous credit banks to enhance the development of the modern business sector. Nyang (1992, p. 429) has warned that so long as donors, development scholars and policy makers, ignore the relevance of cultural, religious and ethnic factors, home-grown development deeply rooted in the history, culture, and psychology of African peoples will be stymied.

Business Ethics and Trust

Islamic and Hausa economic ethics encourage wealth acquisition, so, on face value, they are essentially capitalistic. An in-depth inquiry into Islamic economic thought reveals the centrality of social concerns. One of its fundamental principles is the assumption that economic pursuits must be effectively balanced with concerns for the general community so that wealth can have religious sanctification. These are the fundamental principles that provide guidance to entrepreneurial ethics. Islamic economic doctrines oscillate between the pragmatic aspects of capitalist acquisitiveness and the socialist concern for the community or group. In particular, they eschew the extremism of capitalist exploitation on the one hand and socialist opposition to private property on the other. The basic tenets of entrepreneurial ethics are derived from individual and group values. Entrepreneurial economic ethics as shown in this study, are reinforced by the profit motive and the desire for personal distinction in business activity.

Entrepreneurial rationality or rational behaviour as this study portends, is influenced by the environment and its constraints. Environmental structures and constraints determine eventually the strategies, and options which entrepreneurs may adopt in a particular society. These constraints which were extensively examined include, logistics, history of business performance, reliability of the legal system pertaining to the maintenance of contract law, government economic policies, such as tax regimes, foreign exchange allocations, interest rates etc. Given the unpredictable nature of these institutions, the cautious attitudes of entrepreneurs show that they are rational. Part of rationality of entrepreneurs in modern industry in cost effectiveness. Operators of the three plastic factories of Kano have shown cost effectiveness through a combination of imported raw materials with cheaper recycled products from garbage heaps in the manufacturing process. Thus, entrepreneurial risk perception, and resource control strategies have direct bearings on the way they rationalise their activities.

The strict controls in the Chinese factory rather than creating employee compliance produced the most adverse and hostile behaviour among the workers. This might suggest that managerial flexibility, trust, and empowerment of employees are better alternatives to a regime of stringent controls in the work environment. There was no evidence in any of the three plastic firms that employee attitudes constrained high productivity. To the contrary, both indigenous and expatriate operators of the plastic factories expressed relative satisfaction with the performance of their workers.

The evidence in this study does not provide support to the claim that expatriate business operators, have better ways of resolving employee problems than their indigenous counterparts. Nearly all business firms operating in Nigeria are inescapably dominated by kinship and ethnic networks because of the informal recruitment methods often employed by indigenous managers entrusted with employee selection matters. This method of recruitment constrained employee union activities in the plastic factories.

Today, a passionate criticism of Nigerian business people is for lack of trust (Schatz, 1977, p. 87). The low degree of trust, which is not restricted to one region or a particular ethnic group, has unfortunately become a Nigerian phenomenon. In a survey of 419 western Nigerian business operators on obstacles 'to the formation of partnership or joint ventures', Schatz observes that 75 percent saw such ventures as 'difficult' or 'undesirable because of financial untrustworthiness'. The low degree of trust in Nigerian business circles requires the desire for face-to-face transaction in the Nigerian business sector. This desire for face-to-face transaction inevitably increases transaction costs in terms of expenditure on unwarranted travels, incursion into business time, highway robbery of business travellers and the lost of human lives resulting from motor accidents.

In analysing the dynamics, obstacles, and prospects for building trust relations in Kano and Nigeria as a whole, at least two essential factors have to be taken into account. First is the extent to which cultural forms adapt to or resist new patterns of rationalisation in modern economic activities. And second, the way in which political, economic and ethnic factors intrude upon, hinder, or facilitate the building of entrepreneurial ethics and trust relations in the economic environment. Such analyses are relevant because they provide us with measures or yardsticks on how to integrate indigenous and foreign values in the pursuit of modern economic activity. In addition, they allow us to know the extent to which foreign economic values can be accepted without hampering or obliterating indigenous cultural values.

Paradoxically, there is an inverse trust relation in Kano and especially many Nigerian business circles. The inverse trust is subsumed in the high degree of trust placed in foreign rather than local or coethnic business groups. Foreign partners are trusted on the belief that their respect for indigenous laws makes them more reliable. This belief though may positively encourage foreign participation in the local economic environment constrains the growth of large-scale indigenous corporate ventures in the modern manufacturing sector.

The absence of employee training programmes in the Hausa, Lebanese, and Chinese plastic firms showed that the management of these firms were not future-oriented. It also among other things shows a low level of trust among entrepreneurs in the business climate.

In the plastic sector, entrepreneurs use trusted contacts to increase their own credibility and trustworthiness. In times of raw material scarcity and acute oil shortages, entrepreneurial networks the plastic industry are important in providing information on markets where raw materials or illegal petroleum dealers can be found. Through a network of contacts, plastic manufacturers in Kano can purchase recycled plastic materials from the informal markets of Chad, Cameroon, and the Niger republic. Furthermore, network relations are crucial in providing market information about prices and the performance of competitors. It therefore seems that without networks in the northern Nigerian plastic industry, factory machines will cease to operate.

Religious networks provide avenues for establishing trust or confidence with trading partners. Sometimes, it is a medium for introducing clients to new economic opportunities. In Hausa community, personal piety and attachment to faith are important in providing a kind of security and access to credit facility. Religion is relevant because of the notion that a God-fearing person seldom reneges on his/her promises. Religion facilitates inter-urban marriages. Familial background, class, and economic success often provide conditions for such marriages. Marriage ties in Kano often transform into economic bonds. The *ulama*-merchant alliance in Kano permeates both the religious and economic spheres. In the business sphere, most people provide fronting roles for the *ulama* and ruling classes. Marriage ties with ruling and local *malam* classes may be exploited to boost personnel prestige, reputation, and trust in the urban economy.[270] Clientelistic networks and social contacts are avenues for gaining employment in the modern industrial sector or access to the market. Evidence in this book shows that the general pattern of low level of trust was not entirely cultural. Some of the reasons identified included, uncertainty in the political, economic and social lives of Nigeria.

The panacea for the building of trust in northern Nigerian or Nigerian business is not an easy one because of the interplay of several factors. Trust is not a thing that can be imposed on a person or group; it has to do with changes in personal or group psychological attitudes. At the personal level, trust has to do with moral integrity. The way and manner human beings value integrity and credibility differ from one person to another. Nevertheless, the maintenance or lost of credibility are crucial factors in determining future outcomes in business relations. In the factories, managerial failure to fulfil promises to employee unions or pay negotiated wages and salaries, are often the bases of distrust, which hastens spontaneous or violent reactions.

Building local confidence towards partnership ventures requires efficient legal institutions so that contract agreements and rights in joint businesses can become legally enforceable by an impartial judicial system. The judicial or legal system as presently constituted is incapable of fulfilling such obligations. This implies the building of independent and credible judicial system not only in Kano, but entire Nigerian federation is necessary for laying a solid foundation for trust.

On the ethical level, people must learn to be more reliable or trustworthy in order to attract the confidence of those around them. Government has a role to play in building strong institutions that will forestall trust by curtailing arbitrariness in reward and punishment among the ethic groups.

Increased trust depends on the improvement of societal institutions and proper education of especially local investors about the functioning of the modern economic sector. The leading role of the state in this regard is currently discouraging. Democratic institutions must be built in a way that can raise public, and especially entrepreneurial, confidence about the system in which they operate. The establishment and restoration of public confidence in the judicial system and provision of a legal framework that is capable of protecting individual assets in jointly operated business are indispensable in this regard.

Second, entrepreneurial attitudes towards division of labour need to improve and extend beyond family and primordial circles. Indigenous entrepreneurs must adapt to modern economic division of labour and be prepared to delegate, rather than centralise authority in personal and family hands. This is because the progress of the modern industry thrives on flexibility, division of labour and acceptance of change. Studies have shown that delegation of power to workers is often repaid through trust and hard work.[271] Relatively few Nigerian firms satisfy these requirements. The economic transformations and growth of giant corporations elsewhere suggest that trust is an important factor needed for the expansion of business frontiers. Indigenous Hausa, and indeed Nigerian, entrepreneurs also have to come to terms with the reality that, in modern business transactions skill and competence rather than social backgrounds are the bases for assigning key roles to participants.

Third, the employers of labour in Kano and elsewhere in developing countries have to recognise the fact that an essential condition of modern industry is acceptance of differences and the building of employee trust relations. Industrial or corporate differences are not necessarily a hindrance; their acceptance can be turned into competitive advantage. Employee unions should be regarded as partners in economic progress rather than saboteurs.

Fourth, local attitudes towards time are incompatible with the discipline required for modern business transactions. In order to fit into the requirements of modern manufacturing activity, indigenous Hausa and Nigerian entrepreneurs need to embrace the modern concept of mechanistic time and the idea of punctuality as yardsticks for progress. Entrepreneurs should come to terms with the fact that no meaningful economic growth can be achieved without some appreciable level of technical knowledge. Indigenous employers of labour have to invest in the future by educating and training their employees. Employee training and human resource development is not only a duty of large transnational and multinational corporations, but also an essential condition for the advancement of small indigenous family businesses in the modern industrial sector.

There is no single or moral standard for building enduring trust in society, although educating people is a necessary step. As Luhmann (1979) puts it, trust has to be learned. Yet education alone is not a sufficient condition. A cursory look at Kano and in the Nigerian society shows that the building of trust will go beyond educating the people. Untrustworthy or fraudulent business practices such as the advance fee fraud (AFF) or 419 has a link with the general standard of living. The increasing gaps between the rich and the poor and general unemployment among young people struggling with the vicissitudes of daily life has increased the level of business fraud, violent robbery in recent years. The emergence of city vagrants or miscreants such as the 'Lagos Area Boys' and the Kano *Yan daba*, has a direct relation with Nigeria's economic decline.

Poor remuneration of the Nigerian police force over the years, has not only incapacitated the organisation, it has also increased the vulnerability of travelling business groups to incessant security harassments. Thus, business travellers are as weary of armed robbers as they are of uniformed police officers at road check points. The responsibility of ensuring order, safety of lives and business property is a government duty. Private entrepreneurs have to be assured that police mounting roadblocks are doing so in their interest. Public security is not a personal matter and government functionaries need to readdress the issue of national security with more seriousness. This will entail building an entirely new and professional police force that is also well remunerated.

For trust in the Nigerian economic arena to improve, armed robbery has to be brought under control. In high trust societies, business transaction has advanced beyond the threat of armed robbers. Curtailing armed robbery is not an easy task. The evidence in Nigeria shows that armed robbery cannot be prevented or stopped through the use force. It requires a combination carrot and stick approach. Reducing the income inequality gap through the creation of jobs is the carrot. In the current Nigerian situation, jobs have to be created as a crime prevention strategy. That university graduates remain jobless ten years after graduation only helps to transform desperate well-educated youths into criminals and full-time swindlers. The stick on the other hand implies building formidable and well-equipped law enforcement agents capable of bringing crime under control.

That many business persons keep in their custody large sums of money has to do with the unreliable financial or banking sector. Business groups have to be able to trust that the banks in which they invest their hard-earned incomes will not disappear or be closed overnight. The creation of the Failed Banks Tribunal in 1996, to address the problems of distressed banks is a step in the right direction. Nevertheless, making the banking sector more viable will require more commitment on the part of the Federal Government and private operators of commercial and merchant banks in Nigeria. Trust in the banking sector will rest on the building of viable and reliable financial system.

Foreign investment in Nigeria has witnessed more decline since the late 1980s because of increasing business fraud, corruption, and the inability of succeeding military administrations to deal with business-related types of crime. As the world approaches the next millennium, the main challenges to the Nigerian business sphere will include, building positive entrepreneurial ethics and above all, overcoming the low degree of trust that continues to obstruct the leap towards great economic advancement.

Notes

1 For further details, see especially, Geerts (1993).
2 Culture is considered as distinct from religion in this study, although the two possess common overlapping features in values, norms, belief systems, myths etc. Religion is a more formal or institutionalised expression of belief, with more universal characteristics that often unite people across cultures. In any case, religion transcends cultural or geographical boundaries.
3 For further reading on the capitalist organisation of production through non-capitalist relations, see especially Conti (1979).
4 Actions based on instrumental values provide means of achieving other ends. They are generally designed to increase the actor's stock of 'fungible resources', whereas actions based on immanent values are 'desired purely for their own sake' (Hechter, 1993).
5 See especially, Peacock (1978).
6 In this study, 'modern manufacturing' refers to the sector of the economy characterised by activities carried out with the help of automation and machines driven by non-human power.
7 Ogunbanjo's (1983), study provides a vivid illustration on the way Nigerian workers straddle between local values and modern economic activity involving foreign technology and capitalist modes of production.
8 On the relevance of case studies in organisational research, see Jick (1979), and Faules (1982).
9 The 'why' related questions allow deeper probes into more specific areas. The interviews, which were mostly unstructured, were conducted in Hausa and English and recorded on tape and in a field diary. A few structured questions were added regarding the demographic and social characteristics of respondents.
10 For further reading on the participant observation, see especially Epstein (1967).
11 Secondary data in this usage refers to documented materials from published and unpublished sources.
12 Bryman (1988, p. 102) is particularly of the view that a fusion of interviews, observations, and historical analyses in research allows for a better approach in providing linkages between people and events and 'people's interpretations of the factors which produce such connections'.
13 The false reference to culture by many African leaders as a means of perpetuating bad governance tacitly lends false credence to culture as a negative factor in Africa's development. Ironically, most of the leaders at the forefront of cultural explanations are those who exhibit high levels of disregard for their indigenous cultures in their everyday lives.
14 For further reading, see Marc (1992, p. 255).
15 In 1968, a UNESCO proclamation saw culture as providing the most relevant input to development. However, no practical effort was made to carry this proclamation further until recent years. See for instance Salim (1992).
16 Sethuraman's (1981) study of the Kenyan informal sector has shown that the sector is capable of generating a surplus of capital as much as six to ten times in an average period of five years.

17 The environment is made up of natural and human resources, while the elements of the super structure are the political, religious and cultural patterns that interact with each other to produce the forms of economic and technological activity in every society.
18 On the impact of cultural values and institutional arrangements on the nature of business systems, see especially, Hage (1996, pp. 1–56).
19 In Nigerian firms, bypassing the organisational tier of hierarchy and reaching the boss, which is commonly expressed as 'long-leg', constitutes an expression of closeness with the boss, *uban-gida* or *oga.*
20 The study by Laurent (1983) provides a descriptive background on the cultural diversity of Western conceptions of management.
21 For Betalanffy (1960), equifinality implies that a given state could be attained through several paths or processes.
22 For Hannerz (1995, pp. 31–51) the core of the concept 'Creole' involves 'diversity, interconnectedness and innovation, in the context of global center-periphery relationships'. The 'increasing inter-connectedness of the world', he explains is not necessarily inhibitive but 'also leads to cultural gain'. Though the concept implies mixture and 'constraints of inequality', it does not necessarily connote the total weakening of indigenous cultures. According to Hannerz, part of the creolisation process in Nigeria involves the creation of home-grown religious sects.
23 For an excellent reading on the different roles of the entrepreneur, see Woodward *et al.* (1994).
24 Ibid., p. 7.
25 People with entrepreneurial qualities display high levels of enthusiasm, charisma, creativity, persuasiveness, and tolerance. Thus, founders of successful enterprises usually exhibit high leadership qualities. The entrepreneur is an indispensable medium in any form of economic transformation or development. Innovative entrepreneurs constantly adjust themselves to societal changes. For further reading on entrepreneurial attributes, see Mescom *et al* (1981, p. 26), and Gerber (1985).
26 Schumpeter (1934), for instance, perceives the role of the entrepreneur as lost once a business enterprise has been established.
27 See, for instance, Wilken (1976, p. 60).
28 For further reading on high achievement motives and entrepreneurship, see McClelland (1961, pp. 115–122).
29 See for instance Grégoire (1992, p. 2).
30 Marries (1968), observes that 'no sooner does a man rise above his kinsfolk than imperative demands rob him of profits'. Such demands include unlimited credit facilities, provision of jobs, payment of medical bills, school fees, and so on.
31 On this account, see the Qur-an (*Surat Isra-il*, xvii, v. 27).
32 See Quranic commentary 1975, numbers 6266–2267 by Yusuf Ali.
33 Most scholars consider hedonism an obstacle to African capitalism because, while it encourages acquisition, it discourages accumulation. It is necessary to state, however, that hedonism is not a necessary trait of Hausa culture. However, sometimes the attempts by successful persons to support impoverished kin groups conjure a false picture of extravagance. For further reading on entrepreneurial methods of capital accumulation, see in particular Adam Smith (1981; *[1776, 1789]*), *Wealth of Nations*; Book II, Chapter III.
34 See, for instance, Eicner and Leidholm (1970).
35 Although Gonja is an ethnic group in northern Ghana, in daily Hausa usage, the name has a rather wider application. It is often associated with old transit markets of northern Ghana, including Salaga, Yendi, and Gambaga, from which the important commodity, Ashante colanuts, was bought. Hopkin's (1973) historical narrative is very elaborate on the nature of this trade.
36 For further reading on the development of capitalism in northern Nigeria, see Shenton (1986).

37 Hogendorn's (1978) study provides insights on the role of a cash crop in launching Kano into the world capitalist system and economic prominence over other northern emirates of Nigeria.

38 An account of the somewhat accidental development and dominance of groundnut as a major cash crop and source of revenue in northern Nigeria over the much-favoured cotton by the British government is provided in the Atkins report of 1906.

39 Watts' (1983) vivid account of rising food prices and perennial drought during the peak of cash crop exports is particularly worthy of mention.

40 Smith (1955) and Tahir (1975) provide comprehensive accounts of the methods used by wealthy Hausa merchants to disburse capital to less wealthy traders in their communities.

41 The Hadejia-Jamare Irrigation Project, under the Kano River Project (KRP), a project of approximately 60,000 acres, was created during this period. The deleterious impact of this irrigation project on the natural ecology and cost to small peasant holdings is well articulated by Lubeck (1986, p. 51).

42 For more on this, see (Mustapha 1985, p. 46; and Lubeck, 1986, p. 53).

43 For further reading, see Olukoshi (1986).

44 For a detailed account on the different measures in capital analysis, see for instance, Ward (1976, p. 15).

45 See, for instance, Armer and Armer (1977).

46 See especially Marx in *Capital* III, p.384.

47 On convertible forms of capital, see (Calhoun, 1995).

48 In recent years, floundering financial institutions and long queues in Nigerian banks have given a boost to a variety of traditional saving schemes. In the Yoruba society, for instance, a traditional savings scheme known as *Alajo* is very popular. The system, which is close to the Hausa *asusu,* is a traditional mobile bank. Those who operate this scheme are sometimes called the 'biking bankers,' for the simple reason that they conduct business on bicycles. The *Alajos,* who operate an interest-free savings scheme, normally charge their customers the equivalent of one day's savings a month. Although the persons who operate this scheme are older and well-trusted persons, sometimes *Alajos* may disappear with substantial savings from a community, throwing the entire business community in disarray.

49 For further reading, pertaining to entrepreneurial relations with foreign capital in the context of Federal Government's indigenisation policies, see, Hoogvelt (1979).

50 Beckman is of the opinions that local policies did not actually transform the 'subordination' of indigenous entrepreneurs to an 'extended penetration' into the capitalist world economy.

51 A common indicator of entrepreneurial or managerial knowledge of the market in which they operate is their extent of 'alertness' to market forces. The importance of 'alertness' in business, particularly in commerce, is that, it enables 'smart' business individuals to exploit the ignorance of consumers to make quick profits.

52 See Levine (1966).

53 Oshagbemi (1983) observes that some of the common observable traits among successful entrepreneurs include, a strong desire for independence, the will to succeed amidst all odds, and a practical outlook and good skill in human relations.

54 The coercive impact of colonial rule in restructuring the pre-colonial economy of Kano is well documented in the study of Bello (1982).

55 For further reading, see Mustapha (1985, p. 24–53).

56 The study by Jaggar (1973) on 'Kano City blacksmiths' provides a particularly illuminating illustration of the pre-colonial importance of the craft industry.

57 For a detailed account on the craft industry in northern Nigeria, see Smith (1981).

58 Little research exists on craft activity in Nigeria. Statistics are either lacking or inadequate. Koll's (1969) study of craft activities in the western part of the country revealed a lack of 'empirical research on capital formations in the craft sector'.

59 Throughout sub-Saharan Africa, the quests by indigenous craft owners to protect ancestral professions and skills have resulted in an inability to attract support from local wealthy

business groups. The study by Shea (1974) on the dying industry in pre-colonial Kano, is very elaborate on this problem.

60 For further reading on cultural influence on perception, see Allport and Pettigrew (1957).

61 For a readable account on the distinction between cognition and perception, see Keesing (1974, pp. 73–79).

62 In risk measurement, there are two approaches: objective and subjective. The objective approach regards probabilities as irrelevant in risk analysis (Thornhill, 1989). The objective approach emphasises 'the decision-maker's experience, intuition, and judgement. In this context, a considerable amount of options are eliminated' and only the most likely to succeed are adopted. In the subjective approach, the focus is on all probability factors. Each option is weighed on probabilities and the least risky option is adopted.

63 In a study of Intrinsic Risk Attitude (IRA) derived from the ranking of gamblers, Raiffa (1968), observes the innate and ingrained characteristics in certain people make them less risk-averse than others.

64 See Carlson and Davies (1971).

65 In Hausa linguistic terms, there are a variety of terms and euphemistic epitaphs associated with gambling. A common Hausa saying is that 'gambling, or *caca,* estranges a man even if he has up to a hundred thousand kinsmen'.

66 The Qur-an (*Surat Baqara*, V. 219; and *Surat Ma'ida*, 93), for instance maintains that the 'sin of gambling is greater than profit'.

67 The Federal ministry of Industry and Commerce of Nigeria defines 'small and medium scale firms' as those with total investment of between one hundred thousand naira (equivalent to $1,219), and two million naira ($24,390). This is exclusive of land, but includes working capital. See the *Industrial Policy of Nigeria* (1988).

68 For an excellent reading on 'safety-first' strategy regarding assets in risk-taking, see Hicks (1962).

69 Strategic perspectives constitute ten schools of thought (Mintzberg, 1985, pp. 3–9). This study combines three strategic schools. These include planning, entrepreneurial, and cultural. In the planning school, 'strategy formation is a systematic formal process focusing on planning procedures'. The entrepreneurial school considers strategy formation dependent on the vision of the entrepreneur. On the other hand, the cultural school perceives strategy formation as 'dependent on collective value systems' both external and internal to the organisation. See for instance (Haahti, 1989 Chapter 2). Gupta and Strengrevics (1983) have stressed the importance of organisational culture in shaping strategy implementation in business enterprises. Entrepreneurial strategies are therefore dynamic phenomena; their nature and application depend on the stage of societal development (Ray, 1988, p. 315).

70 For a detailed account, see Forrest (1994, pp. 205–218).

71 For further details, see, Table A.1 in the Appendix.

72 For further details on the Muslim population in Kano and other parts of Nigeria, see Table A.2 in the Appendix.

73 The present-day Kano State officially came into being on May 27, 1967. Until recently, Kano has been ranked as Nigeria's most populous State. The creation of Jigawa state out of Kano in 1991, however, led to a decline in its original population. See for instance, the *National Population Commission's (NPC) Report* (1995).

74 There are varieties of ethnic stereotypes in Nigeria. The projection of the native of Kano as a person with a penchant for commerce and an insatiable taste for money is one of the most commonly known stereotypes.

75 The term 'class' is used throughout this study in a non-stringent occupational context.

76 For further reading on mistrust in kinship economic relations see, Tahir (1975) and Griffith (1987).

77 On the subject of trust and size of family businesses, see (Adler, 1991; Fukuyama, 1996).

78 For further reading, see Tahir (1975).

79 The authorship and source of the *Kano Chronicle*; a document of Hausa History is very

controversial. The earlier versions of the *Chronicle*, which were written in original and fluent Arabic, contrast sharply with those that followed later in *ajami*. Smith (1983), suggests that the small communities of Ghadames Arab who first settled in Kano between AD 1452–1463 perhaps wrote the earlier versions, while the *ajami* versions could be traced to Nigerian authors. In contrast to Smith, Hisket (1957), who traces the source of three Arabic versions of the *Chronicle* to a single Kano author, holds a contrary view. Despite the series of anachronisms and discrepancies in the dates found in these documents, they perhaps still remain one of the most authentic sources of Kano history.

80 For further reading on Islam in northern Nigeria, see the various studies by Steed (1995), Falola and Adebayo (1985).

81 There is now a legal thought or *fatwa* in parts of the Islamic world, which permits a limited charging of interest rates on loans. The school, which draws its authority from the Qur'an, argues that prohibitive usury or the doubling of interest rates on loans is what Islam condemns and not its total ban. In countries such as Indonesia, Turkey, and Malaysia, banks charge minimal interest rates. Despite these isolated exceptions, the position of the Qur'an regarding the charging of interest rates on loans as a social vice has become a *doxa* in the Islamic world.

82 Some of the economic activities that contradict Islamic beliefs and carry a strong stigma include the attachment of interest rates on loans and the manufacture and sale of alcoholic products. Other economic activities regarded as offensive include advertising with Islamic inscriptions on textile prints, carpets, and sleeping mats.

83 On the place of mosques in the Islamic community, see Zaharandeen (1983).

84 The term *Sufi*, which comes from the Arabic word *suf*, refers to a coarse garment in which early Muslim ascetics clothed themselves. For more details, see (Trimingham, 1971, p. 1).

85 The appellations *Tijaniyya* and *Qadiriyya* derive from the names of the 'founders' of the religious sects. In spite of the lack of accurate figures on religious affiliations, it is widely believed that the *Tijaniyya* have the most adherents in Kano.

86 Unlike *Sunnis*, the *Sufis* venerate saints, or *Waliyai*, and the living descendants of the Prophet, known as the *Sharifai*. Among the *Tijaniyya*, the *Salat-al-Fatih*, *Jauharat al-kamal,* and the *Jawahir al Ma' ani* of Sheikh Al Abbas al-Tijani are the most consulted and recited texts. With the *Qadirriya*, the classical text of Sheikh Abdul Qadiri Gilani, the *Futul al gaib* is held in high esteem. Other Islamic texts which are widely consulted by the *Sufis* are essentially *fiqh* literature which include: the *Risalat* of Ibn Abu Zaid-Qairawan, *Mukhtasar* of Sidi Khalil, *Al-Muwatta* of Imam Malik, *Madhal* of Ibn Al-Hajj, *Tuhfat al-Hukam* of Ibn Asim, *Akdari* of Abu Zayd Abdul-Rahman, the *Ashmawi* of Sheikh Abdul-Bari al-Ashmawi ar Rifa-i, and the *Di-wan* and *Mimiyyat Ibrahimiyya* of the late Sheikh Ibrahim Niasse of Kaolak, and the various works of Sheikh Nasiru Kabara and Sheikh Dahiru Bauchi. Most of the hagiographic literature of *Sufi* scholars in Nigeria are focused on the fundamental questions of faith and religious worship.

87 For further details, see Imo (1995, pp. 15–69).

88 The five pillars of Islam include the proclamation of faith, or *tauhid*; the observance of five daily prayers, or *sallat*; fasting in the month of *Ramaddan*; payment of the annual tithe, or *zakkat;* and performing the pilgrimage to Mecca, or *hajj,* by those who have the means.

89 On this account, see Tahir (1975).

90 The religious philanthropic activities of merchant industrialists of Kano such as Aminu Dan-Tata, Issiyaka Rabiu, and Sani Marshal in the construction of mosques and Islamic schools and hosting visiting *Tijaniyya* and *Qadiriyya* scholars are particularly worthy of mention.

91 In Hausa terminology, the monthly wage or *albashi* sometimes literally translates as debtor. This stigma further makes wage employment less attractive.

92 The Hausa term for an industrial worker is *lebura,* from the English noun labourer. Industrial employees in the non-managerial cadre are commonly referred to as *leburori* (plural). This concept slightly differs from *kwadago,* a term usually applied to workers who depend on farm wage.

93 See Adesina (1994).
94 For a comprehensive reading on this particular argument see Turner (1974, p. 7).
95 See for instance Geertz (1963; Turner, 1974; Rodinson, 1980; and Peacock, 1978).
96 For further reading regarding the general misunderstanding of Weber in the academia, see especially (Collins, 1987).
97 Keith Stump's (1984) half century analysis of the growth of world religions strongly debunks the notion of the 'Arab warrior' as a main 'social carrier of Islam'. The study further shows the main ingredients and necessary conditions for the rapid growth of the world's major religions have been peace and stability, rather violence.
98 See especially, Sandbrook and Barker (1985), in their book titled *The Politics of Africa's Economic Stagnation* Cambridge University Press.
99 For an extensive and excellent reading on economic reforms especially regarding 'Muslim Puritans', in Southeast Asia, see Peacock (1973). See also Harris (1986), *The End of the Third World.*
100 For further reading on the economies of Islamic societies, see especially Geertz (1963), *Peddlers and Princes*. See also Rodinson (1980), *Islam and Capitalism.*
101 Bourdieu (1972) for instance, considers 'habitus' as providing the source of the series of moves and strategies that govern rationality in every society. What is rational, logical, or reasonable in one culture may seem non-rational in another. Rationality, is therefore context dependent.
102 For further reading, see Turner (1974, p. 152).
103 In analysing rationality of economic systems, Godelier (1972) examines three categories of actors in the economy, which include entrepreneurs, workers, and consumers as a basis for understanding rationality.
104 Jibrin Ibrahim's (1989) study on the religious groups in northern Nigeria is particularly illuminating.
105 Among the *Izala*, the works of prominent Islamic scholars such as Imam Buhari, Muslim, Tirmidhi, Abu Dawud, Ibn Ma-ja, Jamia-al Sagir and the works of Sheikh Abubakar Mahmud Gumi (1970), the *Akidat al Sahiha*, have gained considerable currency. The *Izala*, which is one of the fastest growing religious movements in Nigeria, encourages learning as a means of avoiding religious pitfalls.
106 The four legal schools recognised in most Islamic societies include the *Hanbali*, *Hanafi*, *Shafi'i*, and *Maliki schools*. In the Maghreb and West African sub-regions, including northern Nigeria, the *Maliki* legal thought predominates. Some extensive and evocative explanations on the Islamic forms of economic transactions appear in the narratives of the *Al-Muwatta* of Imam Malik and the *Sahih al-Bukhari*.
107 For further reading, see Bourdieu (1972, p. 175).
108 A Hausa aphorism that 'a successful businessman is like a king', further buttresses this point. In Kano, many successful persons in business have acquired honorific or chieftaincy titles.
109 See Turner (1974, p. 12).
110 See especially Peacock (1978, p. 136).
111 For further reading, see especially Tahir (1975).
112 A typical characteristic of the bazaar economy is 'one in which the total flow of commerce is fragmented into a very great number of unrelated person to person transactions'. On the reconstruction of the bazaar to a firm economy in developing countries, see Geertz (1980, pp. 28–51).
113 The six key elements of Islamic faith, or *iman,* are the belief in Allah or God, His angels, prophets, His revealed books, the day of resurrection, and in divine preordainment or destiny. In Islamic societies, any aspect of modernisation that encroaches on these basic pillars of faith is considered antithetical to Islam.
114 For further reading on the cultural consequence to local development see especially Nyang (1992, pp. 429–446).
115 Huemer's (1998) study on trust in business relations provides some relevance to this analysis.

116 For further reading on Islamic views regarding trust, or *amana*, see the Qur'an (*Surat Anfal* viii, v. 27, and *Ahzab* xxxiii, V. 72). In these Chapters and verses, very extensive and elaborate explanations regarding trust are provided.
117 Reciprocity in trust relations implies that a demonstration of trust by an individual evokes similar response on the part of another. The studies by Luhmann, 1979; Huemer, 1998; and Dahl, 1998, show that reciprocity is necessary for the maintenance of trust.
118 See especially, Sanner (1997, p. 60).
119 The literature on trust shows that it has a transient character. Thus, whether a trust relation will endure or not depends on the situation. On the transformation and changing forms of trust and distrust, see especially Luhmann (1979, p. 89).
120 On social stigma and long-lasting image created by deviant behaviour, see especially Goffman (1968).
121 See Huemer (1998, p. 131).
122 On competence as a basis of trust, Barber (1983, p. 14) asserts that in complex societies people with technical expertise and professional workers are often regarded as more trustworthy than their less competent counterparts.
123 Most prominent business persons in Kano are well connected with ruling and *malam* classes through marriage. For further reading on marriage ties and the economy of Kano, see especially (Paden, 1973; Tahir, 1975, and Lubeck, 1986).
124 Tahir (1975, p. 260) observes that *ulama* in Kano 'were necessary to all classes and mediated relations between them'. Thus, 'they sold their services to the ruling classes as requested', and 'converted their rewards into capital; ...according to personal choice'. To the merchant, they did the same in teaching him ethic of religion and of commerce.
125 On personal image and self-presentation as instruments of trust, Luhmann (1979, p. 40), is particularly of the opinion that a person's mere appearance presumes some minimum amount of trust. A change in the conditions and prospects of self-presentation may lead to a change of trust. In northern Nigeria as elsewhere in the developing world, personal image is an important yardstick for assessing the capability of individuals. Sometimes a hard-working and assiduous entrepreneur with a poor personality may be denied access to credit facilities whereas a dubious characters with a deceitful personality may be entrusted with big financial loans.
126 See for example, Luhmann (1977, p. 64).
127 For more details see, for instance, Fukuyama (1996).
128 For further reading, see the *Economist Intelligence Unit* (EIU), Report (1994).
129 The Central Bank of Nigeria's annual report for 1994 revealed a widening of the government deficit from 884 million US dollars to 2.27 billion US dollars, which was equivalent to 5.4 percent of the Gross Domestic Product (GDP). The general inflation in the economy meant a sharp decline in the purchasing power of the citizenry to 40 percent of its value in 1987. See, for instance, the *World Bank: May Report* 1994. See also *the Central Bank of Nigeria's (CBN), Annual Report* 1994 and the April/June *Quarterly Report* from the same source.
130 For a detailed account on official monetary policies and their effects on the Nigerian economy, see the *World Bank Report:* May 13 1994. No. 13053-UNI.
131 See the Manufacturing Association of Nigeria (MAN, 1994, p. 10), Half Yearly Economic Review January–June.
132 The Shell International Oil Company, Nigeria's major partner in oil exploration, is currently planning a multi-billion dollar petro-chemical and liquefied natural gas project at Eleme. When completed, the project will provide some of the basic synthetic raw materials needed in the plastic industry. It is, however, unknown whether the current socio-political turmoil in the Niger-Delta will make the realisation of this objective possible.
133 For further details on the raw material situation for various industries, see table A.3 in the Appendix.
134 Garbage picking is a worldwide phenomenon synonymous with the hopelessly poor seeking a means of survival. People seeking to survive under severe economic pressures often use

this very unattractive profession as a last resort. Because of their industrial importance, metal and plastic scraps are among the highly priced items from garbage heaps. For further reading on the garbage industry, see the study by Birbeck (1979).

135 On the economic activities of child entrepreneurs in the urban Kano society, see for instance Schildkrout (1986, p. 195–223).

136 The figures indicate that none of the ten listed industrial sectors was producing up to 50 percent of its capacity, see especially table A.4 in the Appendix.

137 The prominence of foreign investors in the 1970s gave rise to the introduction of the 1972 and 1977 indigenisation decrees. The decrees intended to provide at least 40 percent equity participation of Nigerian business persons in foreign firms operating in the country. There were, however, provisions for 100 percent ownership of expatriate investors who sought to naturalise as Nigerian or African citizens. The first decree, introduced in 1972 and amended in 1977, offered citizenship to foreigners who had lived in Nigeria for at least fifteen years, could speak at least one Nigerian language, and had clean tax and criminal records. This enabled most Levantine and Asian investors to have full control of their businesses. It is, however, interesting to note that the new '*Industrial Policy of Nigeria*' (1988), unconditionally allows foreign investors 100 percent ownership and control of their business, except in such areas as petroleum prospecting, mining, banking, and insurance.

138 The ease with which Lebanese settlers master the indigenous languages distinguishes them from the rest of the expatriate community in Kano and Nigeria in general. A migrant Lebanese businessman, George Galil, is reputed for his ability to speak fourteen local and foreign languages. On this account, see Forrest (1994).

139 For further reading on the Nigerian legal provisions on employee dismissal, see Emiola (1982).

140 For additional information on the human factor in the life of business enterprises, see Waterman (1987).

141 Table A.5 in the Appendix illustrates the channels through which employees in the three plastic firms were recruited.

142 Employee recruitment through social networks was common in all the factories because the people entrusted with recruitment exercises in all the firms were indigenous managers who often had intimate knowledge of the society in which they operated.

143 A job description is a document that clearly states the functions or tasks of an employee and provides a description on how to perform them. The difference between what is described on paper and the employee's performance is known as the performance gap. Through the performance gap, employers or owners of business enterprises can know in what specific area an employee needs further training. In many Nigerian firms, employees are rarely provided with job descriptions because of the poor education and nonchalant attitudes of many employers.

144 See the study by Beatrice Webb (1891).

145 The First reason for collective bargaining is to eliminate disparities in wages for workers who perform identical jobs (Olson, 1965). This is because wage differences become 'a measuring rod' and employees tend to 'establish a dividing line between a square deal and a raw deal' (Ross, 1966, p. 51). Second, collective bargaining unites workers and enables them to achieve goals which would have been impossible if confronted on individual basis (Henley, 1989). A major relevance of collective bargaining is the provision of a medium through which institutionalised dialogue is possible.

146 For further reading see the *I.L.O: World Labour Report* 2, 1985.

147 The proliferation of wage theories from economic and management perspectives often complicate analysis on wage matters. The Industrial Relations approach focuses on political and managerial processes in collective bargaining and their eventual impact in determining wages. This approach does not ignore the relevance of the classical and neo-classical economic approaches. From an economic perspective, supply and demand determine wages, with the individual as the basic unit of analysis. In view of this, changes in market conditions were determinants of relative wages (Brown, and Nolan, 1988). Although wages are by

nature market-sensitive, there are other important factors, such as political policies, government attitudes, unemployment rates, the strength of trade unions, and cultural attitudes towards wage issues.

148 Opsahl and Dunette (1966) argue that wages act as 'prime reinforcers' to the worker because of their association with the basics of life such as food, shelter and clothing. Wages, they observe, provide 'incentives' to the individual when perceived as capable of satisfying various needs. Wages in general act as 'anxiety reducing' agents, particularly in capitalist or competitive societies.

149 See Henley (1989, p. 296).

150 For further reading, see Weeks (1968).

151 See, for example, Argyris (1957, pp. 76–122).

152 For an excellent description on the labour process, see especially Braverman (1974).

153 See, for instance, Burawoy's (1982) study.

154 Table A.6 in the appendix represents selected demographic features of employees in the three plastic firms, which include religious affiliations and ethnic background.

155 For more on the Keynesian view on profit, see, for instance, the review by Keirstead (1961, pp. 547–552).

156 See the various works of Walras (1984; [1874, and 1954]), Schumpeter (1962; [1934), and Brown (1949, [1936]).

157 See Ricardo (1951, and 1955) *Works and Correspondence* edited by Sraffa.

158 The introduction of colonial rule and a more organised cash economy in 1903 has led to the introduction of new ways of calculating profit. Consequently, profits are now calculated on the basis of financial statements on balance sheets. In recent years, the majority of micro-enterprises, which do not have full-time accountants, employ the services of chartered accounting firms to calculate their quarterly or annual profits.

159 For further illustration on entrepreneurial perceptions about profit, see, Table A.7 in the appendix.

160 See Harper 1984:15.

161 For a detailed account, see especially Tahir (1975).

162 See Lubeck (1983, p. 150).

163 Amin's (1967) critique of capitalism in the Ivory Coast provides some useful examples.

164 For further reading, see especially Olukoshi (1985).

165 See Bell (1978).

166 The *American Financial Times Magazine* survey report in 1989 showed that one out every eight firms listed in the FTSEs largest hundred companies have strong family connections. Similarly, the huge Chaebol conglomerates, which have been largely responsible for Korean industrialisation in the contemporary era, are family businesses (Amsden, 1989). For further reading on the family firm, see Church (1993).

167 The honorific *Alhaji* (singular) is the local title of a Muslim who has made the pilgrimage to Mecca. *Alhazai* (Plural), implies many men with such titles. Grégoire (1992) observes that the title *Alhaji* has become a symbol for economic success in Hausa society. The *Alhazai,* usually business proprietors in Kano, are far more prosperous than industrial workers.

168 In organisational studies, a distinction is sometimes made between the terms 'control' and 'controls'. For Drucker (1964) controls as a means may not necessarily lead to the overall objective of control method. Rather, a set of more or less integrated controls may be regarded as a control system.

169 Mockler (1972, p. 2) defines a control measure as 'systematic effect to set performance standards with planning objectives information feedback system to compare actual performance with these predetermined standards, to determine whether there are any deviations, and to measure significance, and to take action required to assure that all corporate resources are used in the most effective and efficient way possible'.

170 For further reading on the relevance of formal controls in business enterprises, see, for instance, Stoner and Freeman (1989).

171 See especially Nafziger (1967), Kilby (1969), and Schatz (1977).
172 See Nancy Adler's (1991) description on types of diversity in groups. The homogenous group, for instance comprises of members of the same culture. A token group, on the other hand, is a homogenous group with one member from a different culture. A bicultural group, as the name implies, has members representing two distinct cultures, whereas a multicultural group comprises at least three or more different cultures.
173 Newman (1975, pp. 6–9) cautions that because of the diversity of activities in modern organisations, no single unified method of control is desirable. This implies that different control methods may be used in a business organisation.
174 Although the Nigerian labour law stipulates that the length of work on a normal day should not exceed eight hours, employers of labour in the country constantly violate this aspect of the law. For further reading on labour provisions regarding working hours in Nigeria, see Emiola (1982, pp. 55–58).
175 On the provisions of the Nigerian labour law regarding work on weekends, see Emiola (1982).
176 Auditing, for Maut and Sharaf (1974) involves the verification and examination of financial data or records for judging the faithfulness with which they portray events and conditions.
177 See the *Plastic Industry Magazine* (1987).
178 For further reading, see especially Barbash (1980).
179 The 1938 trade union ordinance that gave recognition to the formation of trade unions in Nigeria was very flexible. It permitted as few as five employees to form a union, which gave rise to the proliferation of very weak unions in the country. Until 1973, there was no standard rule governing the size of unions in Nigeria. For further reading, see Fashoyin (1980).
180 A national or industrial union refers to 'a trade union whose membership is organised on the basis of representing all grades of workers in a particular industry, and hence includes unskilled, skilled and sometimes non-manual employees' (Marsh, 1979, p. 154). An industrial union can be further described as 'a federation of house unions, crisscrossing several industries' (Otobo, 1987, p. 61). In Nigeria, the industrial unions are most relevant in the collective bargaining process.
181 See in particular the 'Wages of Labour'. This is first in the series of the *Economic and Philosophical Manuscripts* of Karl Marx, written between April and August, 1844 and reproduced in 1971.
182 For further details on legal procedures surrounding dispute settlement in Nigerian firms, see especially Ogunniyi (1991).
183 The term sector refers to an area of the economy where a specific type of activity carried by people subject to the same motives dominates. For further reading, see especially, Rodinson (1980, pp. 54–56).
184 From a Marxist perspective, Moser (1976) provides a vivid illustration on the activities in the informal sector and questions whether this sector is merely dualist or a product of exploitation in urban development of the third world. Similarly, Tokman's (1978) study provides useful illustrations on the links between the informal and formal sectors.
185 See especially Table A.8.
186 The lumpenproletariat in Marxist theorising applies to an amorphous urban social group below the proletariat, consisting of vagrants, tramps, prostitutes, pimps, and beggars. The proletariat, according to Marx, are the wage earners in capitalist societies whose only possession of significant value is their labour. The term lumpenproletariat has been further popularised by Fanon (1966).
187 Hill's (1977, p. 146) study of migration in a Hausa community near Kano, shows that approximately 200 men had migrated compared with 717 who were at home.
188 For further reading on contemporary migration problems in Africa, see especially Baker and Aina (eds.), (1995).
189 A World Bank (1984) report observes that per-capita food production has failed to keep pace with increasing human numbers in most of African countries, while a few studies have indi-

cated slight growth in this area. Some academics claim that rapid population growth has a positive impact on development due to the demand it creates for economic goods, as well as the search for substitutes and technological transformation. In assessing the benefits of population increases, Simon (1982) is optimistic that the problems often associated with large populations are temporary. It therefore remains to be seen how the population potential of most African countries, including Nigeria, will be fully utilised to positive economic advantage in the coming years.

190 Knowles and Anker (1981) observe that the closer migrants are to the home community, the more likely they are to send remittances. Throughout Africa, migrant remittances are important to the homes left behind in rural areas. In northern Nigeria where the circulatory type of migration is common, a family's willingness allow a youth to migrate during the dry season depends on the outcomes of previous expeditions.

191 In a survey of 297 Hausa migrants in Kano between 1972 and 1974, Mortimore (1989) observes that only 13 percent of them claimed to be landless. Less than three percent claimed they had mortgaged or sold their land in the past two years. Up to 57 percent claimed they had left land in the care of parents while 22 percent had entrusted care of their land to other relations. The study further shows that the links between migrants and properties left at home are quite strong and often provide inducements for the desire to return home.

192 The involuntary type of migration is usually caused by 'push factors' such as natural disasters, political and religious upheavals. For further reading, see Hill and Isiugo-Abanihe (1983).

193 The prolonged dry season in northern Nigeria often results in drought. Farmer and Wigley (1985) provide a distinction among three types of drought commonly afflicting tropical Africa. These include agricultural, ecological, and meteorological droughts. The agricultural type of drought, which is most relevant to this discussion, refers to a shortage of water for crop growth.

194 On the phenomenal number and types of migration in Hausaland, Shea (1983) uses terms such as: *kaura,* which involves plain movement from one locality to another; *fatauci*, which implies long distance trade; *almagjiranci* a form of movement in a quest for Islamic knowledge; or 'peripatetic scholarship'. In most Hausa communities, the *almagjirai*, or migrant students, live on the support of alms from the community. *Hajj* is Islamic pilgrimage, and *dandi* refers to flight for personal reasons, often involving questionable motives such as escape from a community because of a disgraceful or very embarrassing episode in a person's life.

195 See Pittin's (1983) study of female Hausa migrants in Katsina.

196 Osuntogun's (1976) study on the causes and implications of city-ward migration for socioeconomic development in Western Nigeria is particularly relevant to this discourse. Similar studies by Olusanya (1969) and Makinwa (1981) shed additional light on this issue.

197 See especially the FAO report (1984).

198 Amin (1995) argues that the ethnic groups in Africa who do not emigrate are the most economically destitute whereas their relatively mobile neighbours are more prosperous. Thus, the Bassari of Eastern Senegal, who do not emigrate, are among the most destitute in the region, whereas the Serere, who do emigrate, have considerably higher incomes. Similarly, the poor Masaido of Tanzania do not emigrate, whereas the farmers of the 'rich Kilmanjaro' region emigrate. Thus, it could be argued that the relative prosperity of the Hausa in business activities compared to other minority ethnic groups in northern Nigeria, is due to the Hausa's greater mobility. Throughout the West African sub-region, the Hausa in diaspora are to be found in lucrative petty commercial and craft activities in almost every town and village.

199 Collectivist here refers to the predisposition of business proprietors toward joint or partnership economic ventures.

200 For further reading on this, see Waldinger *et al.* (1990).

201 In Hausa communities, local and English terms associated with time are used concurrently.

Such terms include, *Janeru, Febreru*, and *Marris*; from the English months of January, February, and March.

202 See especially Mosley (1995, p. 5).

203 See the Qur-an (15:38 and 38:81).

204 For further reading, See especially, Gardet (1976, p. 206).

205 On the Islamic notions about 'time' as expressed in the Qur-an, See Yusuf A Ali's (1975), *English Commentary and Translation of the Holy Qur'an*: Appendix xiii. p. 1620; No. 5678 to lxx. 4 and lxxvi; 1: ciii. 1–3: No. 6262.

206 The potential time refers to the unknown time within which unforeseen events unfold. For a readable account on the African notion about the 'potential time', see Mbiti (1971).

207 A Hausa aphorism that 'remembrance of past is equivalent to prayers' best sums up the reverence in which the past is held in Hausa culture. Most Hausa sayings relating to time and the future assign its knowledge to the Almighty Allah only.

208 For an excellent reading on traditional and modern forms of time perception, see especially, Inkeles and Smith (1974).

209 See especially (Hill, 1977; Callaway, Boyd and Last, 1985; Pellow, 1991; Coles and Mack, 1991).

210 The Arabic equivalent of the adulterous eye is known as the *zina al ayn.* See for instance AlMunajjed (1997).

211 Muslim and believing women are admonished in the Qur'an (33, v. 59) to cast their outer garments over their persons (when outdoors), in conformity to the norms of religious modesty. The Qur'an further encourages proper female dressing as a way of attracting respect and avoiding molestation.

212 See especially Almunajjed (1997, p. 47).

213 Mohammed's (1985) paper is particularly illuminating on this subject. See especially, 'Women in Nigerian History: Examples from the Borno Empire, Nupeland and Igboland', in *Women in Nigeria Today*, Zed Books Ltd., London, pp. 45–51.

214 For further reading on open market economic activities of Hausa women before the Islamic reform in northern Nigeria see especially, (Barth, 1965; Moody 1967; and Hiskett, 1973).

215 For more on this, see especially the *Hadith Al Nabawi* No. 24.

216 Islam forbids the harsh treatment of women and enjoins men to make provision for their protection and maintenance. On the protection and maintenance of women, see the Qur'an (*Surat Al-Nisaa*, vv. 19 and 34).

217 See also (Fapohunda, 1983, p. 43) on the participation of women in the formal economy in Nigeria.

218 There are parts of the Islamic world today in which female employees are accepted in the crucial areas of health and education. In countries such as Algeria, Malaysia, and Indonesia, female factory employees are separated from their male counterparts in order to eliminate mischievous interference and to enhance productivity. In Kano, the majority of *ulama* sees health and teaching sectors as the only ideal areas for female employment.

219 See especially the (1991) census figures from the Federal Republic of Nigeria.

220 On women's individual property rights in Islam, see the Qur'an (4, v. 32).

221 For further reading on the role of Muslim women such as Aisha and Umm Salim in the battle of Uhud see (Al-Bukhari, 1960, p. 125).

222 Child hawking and begging sometimes reflect the economic status of parents in northern Nigeria. In many affluent homes in the urban centres children engaged in educational activities find less time to engage in hawking activities.

223 For an excellent account on work especially outside the formal market economy see especially Glucksmann (1995, p. 69).

224 In recent times, the practice of sponsoring members of the primordial group and relatives to embark on pilgrimage to the Holy Land in Mecca has become very common among wealthy merchants. It should be stated however, that sometimes such generosity is an extension of business strategy. Usually, the sponsored candidates to *hajj* are used as media through which large amounts of goods are imported without attracting custom duties.

225 See the Qur-an (*Sura*, Shams, v. 9–10).
226 A classification of the degree of success of the three firms is illustrated in table A.9. See for instance the Appendix.
227 On this account, see the Qur'an *(Sura Ra'ad*, v. 53).
228 Until the decentralisation and reactivation of the Corporate Affairs Commission (CAC) in the early 1990s, the registration of new businesses in Nigeria, which had hitherto been a voluntary affair, was conducted at two centers in Lagos and the Federal Capital City of Abuja. The staff entrusted with the enormous task of registering several million indigenous and expatriate businesses was scanty and inadequate. Although a new registration drive is currently in full swing, it appears that, even under the new state of affairs, it may require a few additional years of educating small business operators in order to encourage the registration of businesses.
229 A major inhibition of the secretly conducted businesses is the absence of outside support and the cross-fertilisation of ideas from family members and outsiders when things go wrong.
230 Psychologically, an entrepreneur has to prepare for working long hours, being away from home and family for long periods, eating irregular meals, and receiving irregular incomes. Since production and sales targets are by nature irregular, entrepreneurial incomes are usually irregular. Consequently, individuals who engage in business in order to have regular incomes are sometimes disappointed. For further reading on the psychological preparedness of the entrepreneur in particular, see Oshagbemi (1983).
231 Some of the previous studies on the productivity of African labour have been quite superficial to be taken seriously. See, for instance, De Briey's study (1965).
232 The 1996 annual report from the Inspectorate and Environmental Pollution Section of the Federal Ministry of Commerce, Industry and Corporative saw ecological pollution from local factories as one of its major concerns. The report added that smoke from factories during certain times of the year sometimes reduced visibility to less than 20 meters in major cities such as Lagos, Kano, Kaduna, Enugu, and Port Harcourt.
233 Current estimates of Nigeria's population are based on the 1990 national census figures.
234 See the introductory chapter by Schatz (1977) on the nurturing of indigenous capitalism in Nigeria.
235 For more on Africa's debt servicing and its implications on economic growth, see World Bank Report on sub-Saharan Africa April/June 1995. See also *African Business Journal* January (1996).
236 The *Nigerian Vanguard*, of 3 November 1998, provides some relevant information on Nigeria's external debt situation.
237 For a detailed account, see Central Bank of Nigeria (CBN) Annual Report and Statement of Accounts 1997.
238 See, especially, table A.10 in the Appendix.
239 For further details see table A.11 in the Appendix.
240 For a more elaborate discussion on the role of political institutions and economic performance, see North (1996).
241 Klitgaard (1992).
242 For further reading, see Wildavsky (1992).
243 See North and Fukuyama (1996), respectively.
244 In Islamic literature, corruption is portrayed as a forbidden act or *haram*. The exceptional case under which it may be permissible for an individual to resort to bribery in Islam is if it constitutes the only existing avenue by which a person's legitimate rights can be obtained. Apart from this extreme exception, bribery and corruption are acts strongly discouraged.
245 See for instance *The Washington Post*, November Issue Number 16, 1996. See also *The Economist Journal June* 8–14 1996: 48–50. A Berlin based organisation, the 'Transparency International', has lamented that developing nations with dwindling resources are at the forefront of corruption and resource abuses.
246 See the *Nigerian Sunday Times*, 8 April, 1981.

247 The term '419' derives from section 419 of the *Nigerian Criminal Code* which dealt with offenses relating to financial fraud. This criminal code was later promulgated and became known as Advanced Fee Fraud (AFF) Decree No. 13 of 1995. The method of crime usually involves invitation of foreigners through scam letters with promises of profit margins ranging to several millions of dollars, in contracts which include the lifting of crude oil to the purchase of vehicles and hospital equipment. Most of the request that precede the conclusion of business transaction include: 5 percent remittance fee, 1 percent legal fee, 3 percent economic recovery fund tax, Value added tax etc. Usually, the crime perpetuators pretend to be well connected in the Nigerian society with titles and 'bogus prefixes' ranging from Major General, His Royal Highness, Chief, Doctor, *Alhaji* etc. This year (1998) alone, about 2 million scam or '419' letters from Nigeria were intercepted at the John F. Kennedy International Airport in the United States. For an excellent account see, *Today News Paper*, October 25–31, 1998.

248 See Merton (1957).

249 Chambliss (1973) considers bureaucracies to be the starting-point of corruption, especially in developing countries.

250 A recent investigation of corruption charges of key government personnel traced illegally transferred monies to foreign banks including a financial house based in Beirut. See especially the front page of the *Vanguard*, November, 4, 1998.

251 For further reading on various anti-corruption measures in Nigeria, see Chapter Two of the Criminal Code of the Federation of Nigeria and Lagos (1958), sections 98–106 and 114–116. See also Decree 38 of the abrogated *Corrupt Practices Bureau* (1975). In addition, the *Nigerian Constitution* of 1979 and the *Code of Conduct Bureau of Nigeria* (1979) provide useful insights on this subject.

252 See the 'Vanguard News Paper' of 3 November 1998.

253 See the *Kano Ministry of Commerce and Industry Report* (1995).

254 See, in particular, 'Nigeria Business Information' 10/28/98.

255 On this account, see the *Nigerian Newswatch Magazine*, March, 11, 1996, pp. 16–25.

256 Among various cultures and ethnic groups in Nigeria, Cola-nut or *goro* has several symbolic meanings. Apart from its ritual role, the term is used in another context to imply a token gift or bribery. See especially Paul and Roxana Newman (1979).

257 An illustration of employee experience of corrupt practices in the three plastic is provided in Table A.12. See especially the Appendix.

258 However, it is worth noting that Marx and Engels, in their *Communist Manifesto* (1923 [1848]) paid an eloquent tribute to early capitalism: 'The bourgeoisie during its rule of scarce one hundred years has created more massive and more colossal productive forces than have all preceding generations together. Subjection of nature's forces to man, machinery, application of chemistry of industry and agriculture, steam navigation, railways, electric telegraphs, clearing of whole continents for cultivation, canalisation of rivers, whole populations conjured out of the ground - what earlier century had even presentiment that such productive forces slumbered in the lap of social labour'. It is equally well known that Marx considered later stages of the capitalist mode of production as placing 'fetters upon the development of productive forces'. Nevertheless, it is important to distinguish Marx's own thoughts from the simplistic versions of it peddled by some of his followers.

259 On this entrepreneurial attribute, see for instance Fukuyama (1996, p. 6).

260 Among the early adherents of Islam, the first wife of the Prophet, Khadija, and the first caliph of Islam, Abu Bakr, as-Siddiq, and some of the early disciples were very successful merchants who played significant roles in the Islamic society.

261 For further details, see Marimba Ani's critique of western notions of progress.

262 On the ecology and climate of northern Nigeria, see Mortimore (1989).

263 On the performance of the Kano industrial sector during the past four decades, see Figure A.13.

264 For additional information on the size and volume of the manufacturing activities in Kano,

see, for instance, the *Kano Ministry of Commerce, Industry and Corporative Quarterly Report*, January-March 1995.

265 On the notion of incorporating indigenous cultures with economic liberalism as a mean of stimulating prosperity and growth, see Fukuyama (1996, p. 350).

266 See especially Demol *et al* (1982).

267 Freeman's (1979) assessment of Chinese peasant productivity provides some relevance to this discourse.

268 See especially Denisson (1984).

269 For further reading, see for instance, Dia (1992).

270 Most prominent business persons in Kano are well connected with ruling and *malam* classes through marriage. For further reading on marriage ties and the economy of Kano, see especially (Paden, 1973; Tahir, 1975, and Lubeck, 1986).

271 On the relevance of delegated authority and the high level of trust and productivity among employees, see Kanter (1983) and Fukuyama (1996).

Glossary of Hausa and Arabic Terms

adashe	a Hausa rotating credit scheme
aikin Bature	European or factory work
ajami	any written form which is non-Arabic, even if written in Arabic letters
al-barka	blessing from Allah, saints, *malams*, parents, etc.
albashi	end-of-month salary or allowance paid to an employed worker
Alhazai	a honorific title for Muslims who have undergone pilgrimage to Mecca
aljannu	spirits
alkawari	promise or agreement
almajirai	pupils of a Quranic school or beggars
amana	trustworthiness
asusu	money box or traditional safe for keeping money
attajirai	wealthy persons in trade
baiwan Allah	Allah's natural gift
barori	dependants
bid'a	local religious innovativeness
birni	city, used frequently to refer to locations inside the walled area of major northern Nigerian cities such as Kano, Zaria, and Bauchi
bojuwa	long distance commercial activity involving the sale of goods
bori	the cult of being spirit-possessed, especially among rural Hausa women
buta	kettle or gourd in which water is stored for ablution rituals
caca	gambling
ciniki	marketing, trading or selling of things
cinikin kasada	risky business
dandi	emigrating on questionable reasons
dillalai	brokers

dzikr	the remembrance of God through supplication or chanting of litanies
Eid-el-adha	an annual Islamic festival marking the end of pilgrimage to Mecca during which animals are slaughtered
fatauci	long-distance trade
fatwa	Islamic legal ruling
Fiqh	Islamic legal theory
gaib	future occurrences hidden from men but known to Allah only
gaskiya	truthfulness
gayya	communal labour
goro	cola-nut; used symbolically to refer to token gifts, bribery, or corruption
Hadith	words of the Prophet as recorded by his companions
hajj	pilgrimage to Mecca
hakuri	patience
Hambali	school of Islamic law under the teachings of Ahmad Ibn Hambal
Hanafi	school of Islamic law under the teachings of Imam Abu Hanifa
hankali	sobriety
haram	a forbidden act in Islam
Ikhwan,	Muslim Brothers: a Nigerian *Shi'ite* group
ikon Allah	will or grace of God
iman	religious faith which encompass the belief in God, Prophets, Scriptures, Angels, the day of judgement, and destiny
insha Allah	if God so wills
Izala	a Nigerian *Sunni* group that fights local religious innovativeness and advocates Islamic practices based solely on the teachings of the Qur'an and Hadith
jama'a	community or inhabitants
jana'iza	funeral or death ceremony
jari	working capital or initial money used to commence business
Jum'at	a weekly Friday; afternoon congregational prayers
kaddara	destiny
Kantin-kwari	a popular textile market in the Fagge-Tukudu area of Kano
karamci	generousity
karuwai	women with questionable character or prostitutes
kasada	risk or gamble

kashe mu raba	Hausa term for corruption
kasuwa	market or business place
kasuwanci	the act of marketing, selling or trading
kaura	migration
kawalai	pimps
kayan-jabu	counterfeit or fake commodities
kazamin riba	dirty or illegal profit
koli	peddling of small wares
kudin jabu	counterfeit or fake money
kulle	seclusion of married women
kwadago	farming for wages
kwano	bowl or basin made from clay, plastic, or metal
leburori	industrial workers
lokaci	time
long-leg	a pidgin term which indicates bypassing formal rules to achieve personal goals
malam	a learned man or teacher
Maliki	school of Islamic law under the teachings of Imam Anas Ibn Malik
muridai	aspiring students of learned sheikhs or *malams*
Qadiriyya	a belief which claims to be based on the teachings of Sheikh Abdul Qadiri Gilani
Qur-an	Islamic holy book
Ramadan	Islamic month of fasting and abstaining from social pleasures
rana	sun or day
rashawa	bribe
sa'a	luck; hour or time
sabon gari	new town, usually stranger's quarters
sadakat	charity
sana'a	occupation or craft activity
Shafi'i	school of Islamic law under the teachings of Imam Shafi'i
Sharifai	the great-grandsons of the prophet
Shi'ite	a more radical branch of Islam that advocates *Shi'ite* beliefs and practices
shirk	religious associanism or polytheism
sihr	magic
Sufi	Muslim ascetic who believes in mystism
Sunna	acts or practices of the Prophet recommended for all Muslims

Sunni	one of the main branches of orthodox Islam which acknowledges and adheres to the strict practice of the authority or *Sunna* of the Prophet
talakawa	masses or non-aristocratic classes
ta'wil	interpretation
Tijanniya	belief and practice of the teachings of Sheikh Al-Abass Al-Tijani
ulama	the learned men in an Islamic community
wahdaniyya	proclamation or belief in the unity of God
Waliyai	pious men or saints
WAPA	West Africa Pilgrims Association; now a cinema house and popular currency trafficking centre in Kano
waqt	instant time or period
warware alkawari	breaking a promise
wata	month or moon
watar-Bature	European month or month based on the Christian calendar
Wazifa	group supplication among *Sufi* sects which involves the chanting of litanies in unison
yan cin rani	dry season migrants
Yan Tatsine	followers of the late *malam* Muhammadu Marwa, alias *Mai* Tatsine
yau	today
zabi	choice or decision
zakkat	religious tithe, alms, or charity
zuhudu, zuhd	asceticism
zuhr	Islamic prayer conducted after mid-day.

Appendix

Table A.1. Business Diversification by Three Entrepreneurs

A, An Indigenous Hausa Entrepreneur	• Plastic manufacture • Sale of ceramic products • Transporter of goods • Food and grain dealer • Building and construction • Company share holding • Rental of houses
B, A Lebanese Entrepreneur	• Plastic manufacture • Raw material/chemical imports • Skins and hides exports • Commercial textile dealer
C, Chinese Entrepreneurs	• Plastic manufacture • Iron and steel imports • Sweets and confectionery manufacture • Dealer in industrial chemicals • Textile trade • Hotel and restaurant

Source: Author's field interview, 1995.

Table A.2. Percentage of Muslims by State, 1979

State	Muslims
Sokoto	97.6
Kano	97.4
Borno	93.1
Bauchi	80.6
Kwara	75.2
Oyo	62.4
Niger	59.7
Kaduna	56.4
Ogun	54.3
Lagos	44.3
Gongola	34.1
Plateau	25.7
Ondo	12.3
Benue	11.3
Bendel	7.4
Rivers	0.2
Cross River	0.1
Imo	0.1
Anambra	1.6

Source: Henry Bienen, (1985), *Political Conflict and Economic Change in Nigeria*, Frank Cass, London.

Table A.3. Sources of Raw Materials used in Production by Industrial Branch

	Economic Sub-Sector	Imports as % of Raw Materials
1.	Industrial plastic and rubber	79.4
2.	Wood and wood products	22.4
3.	Nonmetallic minerals	23.5
4.	Textiles and wearing apparel	47.6
5.	Chemicals and pharmaceutical	55.8
6.	Food, beverages, tobacco	34.8
7.	Basic metal and fabricated metal products	50.3
8.	Vehicles	78.2
9.	Electrical machinery	80.7
10.	Paper products	60.1

Source: UNIDO, 1988.

Table A.4. Capacity Utilisation by Industrial Sector

Sector	Capacity Utilisation (%) July–Dec 1993	July–Dec 1994
1. Industrial plastics and rubber	40.8	36.4
2. Textiles, wearing apparel	36.1	40.1
3. Wood and wood products	24.5	82.6
4. Paper products	32.5	21.8
5. Chemicals and pharmaceutical	21.6	27.2
6. Nonmetallic minerals	26.1	32.9
7. Food, beverages, tobacco	39.2	25.9
8. Electrical machinery	21.5	21.5
9. Basic metals and fabricated metal products	19.6	21.4
10. Vehicles and miscellaneous assembly	32.5	13.1

Source: MAN 1995.

Table A.5. Managerial Recruitment Channels for Factory Workers

Recruitment channel	Hausa Firm		Lebanese Firm		Chinese Firm	
	Freq.	%	Freq.	%	Freq.	%
(1) Newspaper advertisements	3	3	5	2	68	15
(2) Consulting firms	0	0	0	0	0	0
(3) Social networks 71	71	59	101	41	247	55
(4) Direct hiring at the factory gate	46	38	141	57	138	30
Total N=	120	100	247	100	453	100

Source: Factory Records and author's field interviews, 1995.

Table A.6. Employee Religious Affiliation by Ethnic Background

	Hausa/ Fulani	Kanuri/ Marghi/ Babur	Gwari/ Birom/ Tiv	Yoruba /Igbira	Edo/Ijo	Ibo/ Calabar/ Ibibio	Total Freq.	Total %
Sufi	54	9	2	7	0	0	72	43
Sunni	16	4	1	1	0	0	22	13
Cath.	0	1	8	1	5	16	31	19
Prot.	0	1	7	4	2	6	20	12
Local Christ. Sect.	0	0	12	6	0	0	18	11
Trad. Rel.	0	0	2	0	0	0	2	1
No Resp.	0	0	0	0	1	0	1	1
Total Freq.	70	15	32	19	8	22	166	100

Source: Author's field interview, 1995.

Table A.7. Entrepreneurial/Managerial Definitions of Profit

Value Label	Frequency	Valid Percentage
1. Profit as a difference between buying and selling price	14	32.0
2. Profit as additional earnings on capital investments	22	50.0
3. Profit as assets exceeding total costs and liabilities	4	9.0
4. Profit as surplus of production	4	9.0
Total	44	100

Source: Author's field interviews, 1995.

Table A.8. Urban Labour Force in the Informal Sector of Selected Countries

Country	Share (%)
Nigeria	50
Ivory Coast	31
Ghana	60
Kenya	44
Senegal	50
Tunisia	34

Source: International Labour Organisation (ILO) Geneva 1981.

Table A.9. Distribution of Degree of Success in Three Plastic Firms

Description of Firm	Degree of success
Indigenous Hausa Firm	Moderately successful
Lebanese Firm	Moderately successful
Chinese Firm	Very successful

Source: Author's field study.

Table A.10. Subregional Growth Rates (percentage p.a) 1993-1997

Sub Region	1993	1994	1995	1996	1997
North Africa	0.5	1.8	1.8	4.4	2.8
West Africa	0.5	2.5	3.4	42	3.7
Central Africa	−9.2	−1.3	5	4.4	3.8
Eastern Africa	2.4	4.5	4.9	4.3	3.5
Southern Africa	1.5	2.5	2.5	3	2.4
Africa	0	2	2.7	4	2.9

Source: ECA Secretariat 1998.

Table A.11. Average Ratings of Obstacles in Africa, from the Highest to Lowest

Obstacles	Ratings
1. Corruption	6.65
2. Tax regulation and/or high taxes	4.64
3. Inadequate supply of infrastructure	4.31
4. Inflation	4.19
5. Financing	4.09
6. Crime and theft	4
7. General uncertainty on costs of regulations	3.84
8. Regulations on trade (export and Import)	3.68
9. Policy instability	3.64
10. Labour regulations	3.46
11. Foreign currency regulations	3.44
12. Safety or environmental regulations	3.34
13. Regulations for starting business	3.25
14. Price control	2.75

Source: World Bank, Washington D.C. 1998.

Table A.12. Employee Experience of Corruption in Three Firms

Value label	Freq.	Valid %
1. Experience of corruption between management and government officials	19	11
2. Experience of corruption between management and union officials	22	13
3. Experience of corruption between different levels of factory workers	49	30
4. Never experienced any act of corruption in the work place	76	46
Total	166	100

Source: Author's field interviews, 1995.

Figure A.13. Nigerian and Expatriate Investment in Industry 1950–1989

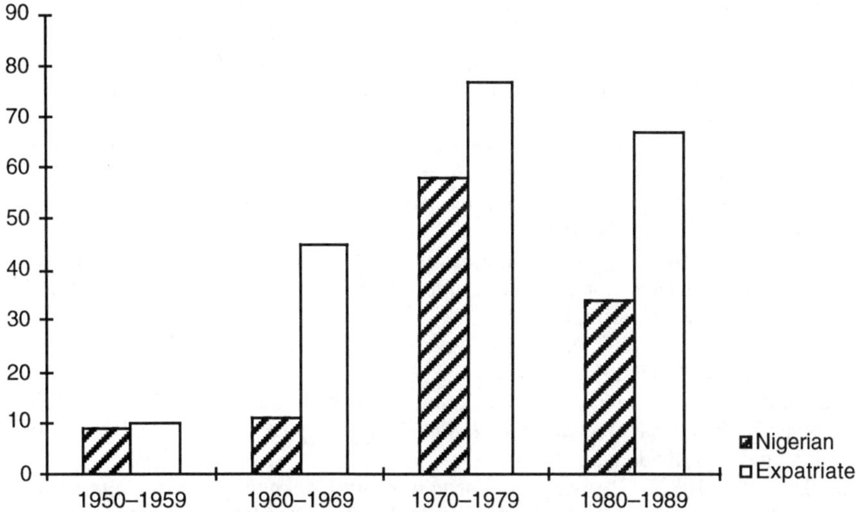

Source: Kano Ministry of Commerce and Industry, 1995.

Bibliography

Aboyade, O. (1972), *The Economics of Bribery and Corruption*, University Press, Ibadan.

Abun-Nasr, J. M. (1965), *The Tijaniyya: A Sufi Order in the Modern World*, Oxford University Press, London.

Adesina, J. O. (1994), 'Unions, Workers and the Development Process', in Himmelstrand, U. *et al.* (eds), *African Perspectives on Development: Controversies, Dilemmas and Openings*, James Curry Publication, London, pp. 183–202.

Adigun, A. (1992), 'Culture, Corruption and Development', *Voices From Africa,* Issue No. 4, pp. 41–52.

Adler, N. J. (1991), *International Dimensions of Organizational Behaviour*: Second Edition, Wadsworth Publishing Company, Belmont California.

Agbor, J. (1990), *Organizational Change Within Cultural Context*, Ph.D Dissertation, University of Lund, Sweden.

Ake, C. (1988), 'Building on the Indigenous', Frühling, P. (ed.), *Recovery in Africa: A Challenge for Development Cooperation in the 1990s*, Swedish Ministry for Foreign Affairs, Stockholm, No. 21.

Akeredolu-Ale, E. O. (1975), *The Underdevelopment of Indigenous Entrepreneurship*, University of Ibadan Press, Ibadan.

Akeredolu-Ale, E. O. (1977), 'The Evolution of Private Indigenous Entrepreneurship in Nigeria', in Teriba, O. *et al.* (eds), *Industrial Development in Nigeria*, University of Ibadan Press, Ibadan, pp. 1–15.

Akinkoye, O. (1982), 'Pathways to Corruption', in Odekunle, F. (ed.), *Nigeria: Corruption in Development*, University of Ibadan Press, Ibadan, pp. 92–95.

Al-Bar, M. A. (1984) *Amaal al-Mar' a fil-l-Mizan,* Ad-dar as Su'udiyya li-n-Nashr, Jeddah.

Al-Bukhari, S. (1987), *The Translations and Meanings of Sahih Al-Buhari*. A Translated Arabic-English (ed.), by Khan, M. M. Vol. 3, Suiwalan, Lahoti Fine Art Press, Delhi.

Ali, Y. A. (1975), *The Holy Qur-an: Text, Translation and Commentary*, Islamic Foundation Publications, London.

Allais, M. (1955), *Fondements d'une Théorie Positive des Choix Comportant une Risque*, Paris.

Allardt, E. (1987), 'Structural, Institutional, and Cultural Explanations', in Himmelstrand, U. (ed.), *The Multiparadigmatic Trend in Sociology*, Almqviell International, Stockholm, pp. 17–33.

Allport, G. W. and Pettigrew, T. F. (1957), 'Cultural Influence on the Perception of

Movement, The Trapezoidal Illusion among the Zulus', *Journal of Abnormal and Social Psychology*, pp.104–113.
AlMunajjed, M. (1997), *Women in Saudi Arabia Today*, Macmillan Press Ltd., Great Britain.
Alvesson, M. (1994), 'Culture as Metaphor, Metaphors for Culture', in Sjögren, A and Janson, L. (eds), *Culture and Management in a Changing Europe*, Institute of International Business, Stockholm, pp. 29–59.
Alvesson, M. (1995), *Cultural Perspectives on Organizations*, Cambridge Press, England.
Amin, S. (1967), *Critique of Capitalism in Ivory Coast*, Editions Minuite, Paris.
Amin, S. (1980), *Class and Nation, Historically and in the Current Crisis*, Monthly Review Press, New York.
Amin, S. (1994), 'The Issue of Democracy in the Contemporary Third World', in Himmelstrand, U. *et al.* (eds), *African Perspectives on Development*, James Curry Publication, London, pp. 320–335.
Amin, S. (1995), 'Migration in Contemporary Africa: A Retrospective View', in Baker, J. and Aina T. A. (eds), *The Migration Experience in Africa*, Gotab Press, Sweden, pp. 29–40.
Amsden, A. H. (1989), *Asia's Next Giant*, Oxford Press, London.
Andersson, T. D. (1987), *Profit in Small Firms: A Model Based on Case Studies from the Informal Sector in a Developing Country*, Gower Company Ltd., England.
Andrae, G. and Beckman, B. (1998), *Union Power in the Nigerian Textile Industry: Labour Regime and Adjustment*, Elanders Gotab, Stockholm.
Ani, M. (1994), *Yurugu: An African-Centered Critique of European Cultural Thought and Behaviour*, Africa World Press, Trenton, New Jersey.
Anikpo, M. (1982), 'Reflections on the Nature of Corruption', in Odekunle, F. (ed.), *Nigeria Corruption in Development*, University of Ibadan Press, Ibadan, pp. 59–65.
Archer, M. S. (1989), *Culture and Agency: The Place of Culture in Social* Theory, Cambridge University Press, Great Britain.
Argyris, C. (1957), *Personality and Organization, the Conflict Between System Individual*, Harper and Brothers, New York.
Atkins, J. C. (1906), *The Story of the British Cotton Growing Association*, Manchester.
Baker, E. L. (1980), *Managing Organization Culture*, Autumn, Mckinsey Quarterly.
Baker, J. and Aina, T. (eds), (1995), *The Migration Experience in Africa*, Gotab Press, Sweden.
Barbash, J. (1980), *Industrial Relations*, IRRA Newsletter, May, Vol. xxii, No. 2.
Barber, B. (1983), *The Logic and Limits of Trust*, Rutgers University Press, New Brunswick, New Jersey.
Barkow, J. (1972), 'Hausa Women and Islam', *Canadian Journal of African Studies*, Vol. 6, No. 2: pp. 317–28.
Barney, J. B and M. H. Hansen (1994), 'Trustworthiness as a Source of Competitive Advantage', *Strategic Management Journal*, No. 15, pp. 23–46.
Barth, H. (1965), *Travels and Discoveries in North and Central Africa*, 1849–1855, Frank Cass, London.
Bashir, I. L. (1986), 'Sources of Capital Formation and Accumulation on Among Kano Entrepreneurs', *African Studies Working Papers*, Boston University Publications, Boston, No. 117.
Baumack, C. (1983), *Basic Small Business Management*, Englewood Cliff, New Jersey.

Bayley, D. H. (1966), 'The Effects of Corruption in a Developing Nation', *Western Political Quarterly,* No. 19.

Beckman, B. (1982), 'Whose State? State and Capitalist Development in Nigeria', *Review of Political Economy,* Sheffield, England, January–April, No. 23, pp. 37–51.

Belasco, B. I. (1980), *The Entrepreneur as Culture Hero*, Praeger Scientific Publications, New York.

Bell, D. (1978), 'The Breakup of Family Capitalism', in Mihai, T. (ed.), *Doctrine Economice Contemporane*, Editura Junimea, Iasi.

Bello, J. A. (1986), *The Changing Dimension of Nigeria's Business Environment*, Macmillan Press, London.

Bello, S. (1982), 'State and Economy in Kano, 1894–1960: A Study of Colonial Domination', *Ph.D Dissertation,* Ahmadu Bello University, Zaria.

Bendix, R. (1962), *Max Weber: An Intellectual Portrait*, Doubleday and Company, Inc., New York.

Berg, E. J. (1968), 'The Development of Labour Force in Sub-Saharan Africa', *Studies in Industrialization,* pp. 350–400.

Berg, E. J. (1969), 'Urban Real Wages and the Nigerian Trade Union Movement, 1939–1960, A Comment', *Economic Development and Cultural Change*, July, xvii, Vol. 4.

Berger, P. L. (1967), 'Religious Institutions' in Smelser, N. J. (eds), *Sociology An Introduction*, New York.

Berger, P. and Luckman I. (1987), *The Social Construction of Reality*, Penguin Books, Middlesex.

Bertalanffy, L. V. (1960), *Problems of Life*, Harper and Row/Torchbook, New York.

Beveridge, A. and Oberschall, A. (eds), (1979), *African Businessmen and Development in Zambia*, N. J., Princeton.

Biersteker, T. J (1978), *Distortions or Development,* The Massachusetts Institute of Technology, USA.

Biersteker, T. J. (1987), 'Indigenization and the African Bourgeoisie: Dependent Development in an African Context', In Lubeck P. M. (ed.), The African Bourgeoisie: *Capitalist Development in Nigeria, Kenya and the Ivory Coast*, Lynne Rienner Publishers, Boulder, Colorado, pp. 249–303.

Birbeck, C. (1979), 'Garbage, Industry and the Vultures of California', in Brombley, R. (ed.), *Casual Work and Poverty in Third World Cities*, John Wiley and Sons Publication, USA.

Boudhiba, A. W. (1982), *La sexualite En Islam*, third edition, Quadrige, Presses Universitaires de France, Paris.

Bourdieu, P. (1972), *Outline of a Social Theory of Practice,* Harvard University Press, Cambridge M. A.

Bourdieu, P. (1991), *Language and Symbolic Power*, (ed.), and introduced by Thompson, J. B. and Translated by Raymond, G. and Adamson, M., Harvard University Press, USA.

Boyd, J. and L. Murray (1985), 'The Role of Women as Agents of Religieux in Sokoto', *Canadian Journal of African Studies*, 19, No, 2, pp. 283–300.

Brachau, R. H. and Nord, W. R. (1979), 'An Exploration of Factors Entrepreneurial Decision: Personnel Management vs Environmental Conditions', *Academy of Management.*

Braverman, H. (1974), *Labour and Monopoly Capital*, Monthly Review Press, New York.

Bromley, A. B. (1993), *Reputation, Image and Impression Management*, John Wiley and Sons, Chichester.

Brown, W. and Nolan P. (1988), 'Wages and Labour Productivity: The Contribution of Industrial Relations to the Understanding Pay Determination', *British Journal of Industrial Relations*, November, Vol. 1–3.

Bryant, C. (1992), 'Culture, Management and Institutional Assessment', in Serageldin, I. and Taborrof, J. (eds), *Culture and Development in Africa*, UNESCO Publications, Washington DC. pp. 447–464.

Bryman, A. (1989), *Research Methods and Organization Studies,* Unwin Hyman, London.

Bulcholz, R. (1977), 'The Belief Structure of Managers Relative Work: Concepts Measured by Factor Analytic Model', *Personnel Psychology*, No. 30, pp. 567–587.

Burawoy, M. (1982), *The Manufacturing Consent*, University of Chicago Press, Chicago.

Burns, T. R. (1997), 'Money in Modern Culture', *Uppsala Theory Circle Seminar Series,* The Department of Sociology University of Uppsala, Sweden.

Burns, T. R. and Dietz, T. (1992), 'Cultural Evolution: Social Rule System, Selection and Human Agency', *International Sociology,* September, Vol. 7 No. 3, pp. 259–283.

Caldwell, J. C. (1976), 'Toward a Restatement of Demographic Transition Theory', *Population and Development Review*, Vol. 2, Nos. 3 and 4.

Calhoun, C. (1995), 'Habitus, Field and Capital: Historical Specifity in the Theory of Practice', *Critical Social Theory*, Blackwell, Cambridge MA, pp. 132–161.

Callaway, B. J. (1987), *Muslim Hausa Women in Nigeria: Tradition and Change,* Syracuse University Press, New York.

Carlson, J. L. and Davies C. M. (1971), 'Cultural Values and the Risk Shift', *Journal of Personality and Social Psychology*, No 20.

Central Bank of Nigeria (1994), *Quarterly Report*, April/June, Lagos.

Central Bank of Nigeria (1994), *Annual Report*, December, Lagos.

Chambliss, W. J. (1973), 'Vice, Corruption, Bureaucracy and Power', in *Sociological Readings in the Conflict Perspective*, in Chambliss, W. J. (ed.), Adison-Wesley.

Chambua, S. E. (1994), 'The Development Debates and Crisis of Development Theories: The Case of Tanzania with Special Emphasis on Peasants, State and Capital', in Himmelstrand U. *et al.* (eds), *African Perspectives on Development*, James Currey, London, pp. 37–50.

Charles, W. (1954), *The History of Unilever*, London.

Chase-Dunn, C. (1981), *Three Approaches to the Kernel of the Capitalist World*, John Hopkins University Press, USA.

Child, J. (1981), 'Culture, Contingency and Capitalism in Cross-National Study of Organizations', in Cummings, L. L. and Staw, B. M. (eds), *Research in Organizational Behaviour*, Vol. 3, JAI Press.

Church, R. (1993), 'The Family in Industrial Capitalism, International Perspective on Hypothesis and History', *Journal of Business History,* October, Vol. 35 No. 4.

Clegg, S. R. (1990), *Modern Organizations: Organization Studies in the Post-Modern World,* Sage Publications, London and Delhi.

Clinard, M. B. and Abbot, D. J. (1973), *Crime in Developing Countries: A Comparative Perspective,* John Wiley and Sons, New York.

Cochran, T. C. (1971), 'The Entrepreneur in Economic Change', in Kilby P. (ed.), *Entrepreneurship and Economic Development*, The Free Press, USA, pp. 95–108.

Cohen, W. A. and Reddick, M. E. (1981), *Successful Marketing for Small Business,* Amacom Publications, New York.

Coleman, D. C. (1987), 'Failings and Achievements of Some British Business: 1910–1980', *Journal of Business History,* Vol. 29.

Coleman, J. C. (1990), *Foundations of Social Theory*, The Belknap Press of Harvad University Press, Cambridge, MA.

Coles, C. (1991), 'Hausa Women's Work in a Declining Urban Economy: Kaduna, Nigeria, 1980–1985', in Coles C. and Mack, B. (ed.), *Hausa Women in the Twentieth Century,* The University of Wisconsin Press, USA, pp. 162–191.

Coles, C. and Mack, B. (eds), (1991), *Hausa Women in the Twentieth Century,* The University of Wisconsin Press, USA.

Collin, L. (1965), 'What is the Problem about Corruption', *Journal of Modern African Studies,* Vol. 3, No. 2.

Coser, L. (1967), *Continuities in the Study of Social Conflict*, Macmillan Free Press, London.

Cottingham, J. (1984), *The Rationalist*, Oxford University Press, London. Cranton, P. (1983), *Proprietary Capitalism: The Textile Manufacture at Philadelphia 1800–1885,* Cambridge.

Cummings, L. L. and Schmidst, S. M. (1972), 'Managerial Attitudes of Greeks; The Role of Culture in Industrialization', *Administrative Quarterly,* No. 17, pp. 265–270.

Cummings, L. L. and Dunham, R. B. (1980), *Introduction to Organizational Behaviour: Text and Readings,* Richard D. Irwin, Homewood Illinois

Czarnawski-Joerges, B. (1987), 'Control Process in Declining Organizations', The Polish Economy 1971–1981, *Organization Studies,* No. 8, Vol. 1, pp. 149–108.

Daft, R. L. and Macintosh N., B. (1984), 'The Nature and Use of Formal Control Systems for Management Control and Strategy Implementation', *Journal of Management,* Fall No.10, Vol. 1.

Dahl, G. (1998), *Trade, Trust, and Networks: Commercial Culture in Late Medieval Italy,* Nordic Academic Press, Sweden.

Dalton, G. W. (1971), *Motivation and Control in Organizations, Irwin, Homewood.*

Dasgupta, P. (1988), 'Trust as a Commodity' in Gambetta, D. (ed.), *Trust:* Making and Breaking Cooperative Relations, Basil Blackwell, New York, pp. 49–72.

Davidson, B. (1984), *The Story of Africa*, Michell Beazelz, London.

Day, B. (1986), 'Meeting Needs, Characteristics of Entrepreneurs: Implications for Education', *Guidance and Counselling,* No 1, Vol. 4, pp. 63–66.

De-Briey, P. (1965), The Productivity of African Labour, *International Labour Review,* August/September, Vol. lxxii, pp. 119–37.

Deedat, A. (1992), *Muhammed the Prophet of Islam*, Islamic Propagation Series, Cape Town.

Demol, E. *et al* (1982), 'Lome's Informal Industrial Sector', in Fransman, M. (ed.), *Industry and Accumulation In Africa*, Heinemann Educational Books, London.

Denison, D. (1984), 'Bringing Corporate Culture to the Bottom Line', *Organizational Dynamics,* Vol. 13, No. 2, pp. 2–22.

Derrida, J. (1991), *A Derrida Reader: Between the Lines*, (ed.), by Kamuf, P., Columbia University Press, New York.

Dia, M. (1992), 'Indigenous Management Practices: Lessons for African Management in the 90s', in Serageldin, I. and Taborrof, J. (eds), *Culture and Development in Africa*, UNESCO Publications, Washington DC, pp. 165–191.

Dobb, M. (1963), *Studies in the Development of Capitalism*, International Publishers, New York.

Douglas, M. (1985), *Risk Acceptability According to Social Sciences: Social Science Research Perspectives,* Occasional Reports on Topics, Russel Sage Foundation, New York.

Drucker, P. (1964), 'Control, Controls and Management', in Bonini, C. P. *et al.* (eds), *Management Controls: New Directions in Basic Research*, Mcgraw-Hill, Maidenhead, pp. 286–96.

Drucker, P. F. (1985), *Tasks, Practices, and Responsibilities*, Abridged and Revised Edition, Harper and Row, New York.

Durkheim, E. (1961 [1912]), *The Elementary Forms of Religious life*, New York. Duvel, W. (1962), 'Small Business Failures', *US Senate Report,* June.

Edwards, E. and Bell, P. (1972), *The Theory and Measurement of Business* Income, Los Angeles.

Eicher, C. K. and Leidholm, C. (eds), (1970), *Growth and Development Economy,* Michigan University Press, USA.

Ekeh, P. P. (1994), 'The Public Realm and Public Finance in Africa', in Himmelstrand, U. *et al.* (eds), *African Perspectives on Development*, James Currey Ltd., London, pp. 234–248.

Emiola, A. (1982), *Nigerian Labour Law,* University of Ibadan Press, Ibadan.

Emory, C. W. and Cooper, D. R. (1991), *Business Research Methods,* fourth edition, Richard D. Irwin Inc., USA.

Epstein, A. L. (1967), *The Craft of Social Anthropology,* Social Science Paper Back, London.

Epstein, C. F. (1970), *Women's Place: Options and Limits in Professional Carriers,* University of California Press, Berkeley.

Essang, S. M. and Mabawonku P. A. (1974), *Determinants and Impact of Rural-Urban Migration, A Case Study of Selected Communities in Western Nigeria,* Michigan University Press, East Anglia.

Etuk, E. J. (1985), *The Nigerian Business Environment: An Introduction to Business Society in an African Economy,* Macmillan Press, Lagos.

Etzioni, A. (1961), *A Comparative Analyses of Complex Organizations*, The Glencoe Free Press, Lagos.

Ewing, A. F. (1968), *Industry in Africa*, Oxford University Press, London.

Eze, N. (1981), 'Motivation and Work Production Amongst Nigerian Worker', *University of Ibadan Seminar Series*, Ibadan.

Fadayomi, T. O. (1988), *Rural Development and Migration in Nigeria*, Nigerian Institute of Social and Economic Research (NISER), Ibadan.

Falola, T. and Adebayo, A. (1985), 'Pre-colonial Nigeria: North of the Niger Benue', in Olaniyan, R. (ed.), *Nigerian History and Culture*, Longman Group (FE), Ltd., Hong Kong, pp. 56–96.

Fama, E. (1980), 'Agency Problems and the Theory of the Firm', *Journal of Political Economy,* April, No. 88, pp. 288–307.

Fanon, F. (1966), *The Wretched of the Earth*, Grove, New York.

FAO Report (1984), 'Migration and Rural Development', in Population Distribution and Development, *Department of International and Social Affairs,* United Nations.

Fapohunda, E. (1976), 'Urban Women's Labour Force Participation Rate Patterns in Nigeria', University of Lagos, *Mimeo.*

Fapohunda, E. (1983), 'Female and Male Work Profiles', in Oppong, C. (ed.), *Female and Male in West Africa*, George Allen and Unwin, London, pp. 32–53.

Farmer, G. and Wigley, T. M. L. (1985), *Climatic Trends for Tropical Africa, A Research Report for the Overseas Development Administration*, Climatic Research Unit, University of East Anglia.

Fashoyin, T. (1980), *Industrial Relations in Nigeria*, Longman Series, Hong Kong.

Faules, D. (1982), 'The Use of Multiple Method in Organizational Setting', *Western Journal of Speech and Communication*, Vol. 46, pp. 150–161.

Federal Ministry of Information (1970), *Second National Development Plan 1970–74,* Lagos.

Fiet, J. O. (1995), 'Risk Avoidance Strategies in Venture Capital Market Economy', *Journal of Management Studies, July* 32:4, pp 551–572.

Flanders, A. (1969), 'The Nature of Collective Bargaining', in Flanders, A. (ed.), *Bargaining*, Penguin series, London.

Forrest, T. (1994), *The Advance of Indigenous Capital in Kano and Kaduna,* Edinburgh Press, London.

Foucault, M. (1984), *The Foucault Reader,* Rabinow, P. (ed.), Pantheon Books, New York.

Frank, A. G. (1969), *Capitalism and Underdevelopment in Latin America*, Penguin Books, Harmondsworth.

Fredrickson, G. (1971), *The Black Image in the White Mind,* Wesleyan University Press, Middletown CT.

Freedman, M. (1979), *Study of Chinese Society,* Stanford University Press, Stanford.

Friedman, J. (1995), 'Global System, Globalization and Parameters of Modernity: Is Modernity a Cultural System'? in Kaarsholm, P. (ed.), From Post-Traditional to Post-Modern? Interpreting the Meaning of Modernity in Third World Urban Societies, International Development Studies Roskilde University, Denmark, Occasional Paper, No. 14, pp. 15–30.

Frijda, N. and Jahoda G. (1966), 'On the Scope and Method of Cross-Cultural Research', *International Journal of Psychology,* Vol. 1. pp. 110–27.

Frishman, A (1991), 'Hausa women in Urban Economy of Kano', in Coles, C. and Mack, B. (eds), *Hausa Women in the Twentieth Century,* The University of Wisconsin Press, USA, pp. 192–203.

Fukuyama, F. (1992), *The End of History and the Last Man,* Free Press, New York.

Fukuyama, F. (1996), *Trust: The Social Virtues and the Creation of Prosperity,* Penguin Books, England.

Fuller, R. B. (1981), *Critical Path,* St. Martin's Press, Washington D. C.

Gano, D. Z. (1983), *Hawking in the City: Needs, Problems and Physical Planning Solutions, A Case Study of Kano,* Unpublished Msc. Thesis, Department of Urban and Regional Planning, Ahmadu Bello University, Zaria.

Gardet, L. (eds), (1976), 'Muslim Views of Time and History', in Gardet, L. *et al,* with an Appendix by Meziane, A., *Culture and Time,* The Unesco Press, Paris, pp. 197–227.

Garlick, P. L. (1971), *African Trader and Economic Development in Ghana,* Clarendon Publication, Oxford.

Gbotokuma, Z. (1992), 'Cultural Identity and Underdevelopment in sub-Saharan Africa', *Voices from Africa Journal*, issue No. 4, pp. 12–28.

Geertz, C. (1963), *Peddlers and Princess, Social Development and Economic Change in Two Indonesian Towns,* University Press, Chicago.

Geertz, C. (1993), *The Interpretation of Cultures*, Fontana Press, London.

Gerber, M. E. (1985), *The Myth: Why Most Business Don't Work and What to Do About it,* Balinger, Cambridge Mass.

Gerholm, L. (1994), 'The Concept of Culture in Ethnology', in Sjögren, A. and Janson, L. (eds), *Culture and Management in a Changing Europe*, Institute of International Business, Stockholm, pp. 15–28.

Ghazali, A. H. al- (1961 [1058–111]), *Ihya Ulum Al-Din*, Dar Al-Kutub, Cairo.

Giddens, A. (1995), *The Consequences of Modernity*, Polity Press, Great Britain.

Glucksmann, M. A. (1985), 'Why Work? Gender and the Total Social Organization of Labour', *Gender and Organization*, April, Vol. 2 No. 2, pp. 63–75.

Goffman, E. (1968), *Stigma*, Penguin Books, Harmondsworth.

Gregoire, E. (1992), *The Alhazai of Maradi: Traditional Hausa Merchants in a Sahelian City,* English Translation by Hardy, B. Lynne Reiner Publishers Inc., Colorado.

Griffith, R. J. *et al.* (1987), 'Variables of Success and Failure in the Third World Development', in Berger, P. L. (ed.), *Modern Capitalism,* Hamilton Press, USA, Vol. 2.

Gugler, J. (1965), 'Life in a Dual System', *East African Institute of Social Research Conference Papers*, January.

Gumi, A. M. (1970), *Akidat Al-Sahihah*, Dar al Kutub, Beirut.

Gupta, A. K and Strengrevics, J. M. (1983), 'Strategy, Culture and Climate: A Conceptual Re-Evaluation', *Boston University, School of Management, Working Paper*, Boston, Vol. 32, No. 83.

Haahti, J. A. (1989), *Entrepreneurs' Strategic Orientation: Modelling Strategic Behaviour in Small Industrial Owner-Managed Firms*, Academiae Oeconomicae Helsingiensis, Acta, Helsinki.

Hage, J. (1996), 'Institutional Context, Choice and Economic Performance', *Uppsala Theory Circle Seminar Paper*, The Department of Sociology, Uppsala University, Sweden.

Haggard, S. (1990), *Pathways from the Periphery*, Cornell University Press, USA.

Hannerz, U. (1995), 'The Social Organization of Creolization', in Kaarsholm, P. (ed.), *From Post-Traditional to Post-Modern?, Interpreting the Meaning of Modernity in Third World Urban Societies*, International Development Studies, Roskilde University, Denmark, Occasional Paper No. 14, pp. 31–51.

Harper, M. (1984), *Small Business in the Third World: Guidelines for Practical Assistance,* John Wiley and Sons, Chichester and New York.

Harris, D. (1971), 'The Nigerian Entrepreneurship in Industry', in Kilby, P. (ed.), *Entrepreneurship and Economic Development*, Michigan University Press, Michigan, pp. 331–55.

Harris, N. (1990), *The End of the Third World: Newly Industrializing Countries and the Declines of an Ideology*, Peguin Books, London.

Hart, K. (1973), 'Informal Income and Opportunities and Structure of Employment in Ghana', *Journal of Modern African Studies,* Vol. 11, pp. 61–89.

Hayashi, S. (1988), *Culture and Management In Japan,* University Press, Tokyo.

Hayek, F. A. (1980a [1948]), *Individualism and Economic Order*, The University of Chicago Press, Chicago.

Hechter, M. (1987), *Principles Of Group Solidarity*, University of California Press, USA.

Hechter, M. *et al* (1993), *The Origin of Values*, Aldine De Gruyter, New York.

Helson, J. (1964), *Adaptation-Level Theory: An experimental and Systematic Approach to Behaviour,* Harper, New York.

Henley, J. S. (1989), 'African Employment Relationships and the Future of Trade Unions', *The British Journal of Industrial Relations,* November, Vol. 27, No. 3.

Herskovits, M. (1952), *Economic Anthropology,* USA.

Hicks, J. (1862), 'Safety-First and the Holding of Assets', *Econometrica* Vol., 4.

High, J. (1982), 'Alertness and Judgement', in Kirzner I. I., *Method, Process, and Austrian Economics*, Lexington Books, Lexington M. A., pp. 161–68.

Hill, A. and Isiogu-Abanihe, U. (1983), 'The Demographic Situation in sub-Saharan Africa: A Background Paper', *The World Bank: Population Health and Nutrition Department,* Washington DC.

Hill, P. (1972), *Rural Hausa: A Village and a Setting,* Cambridge University Press, England.

Hill, P. (1982), *Dry Grain Farming Families,* Cambridge University Press, England.

Himmelstrand, U. (1986), 'Introduction Comments', in Himmelstrand, U. (ed.), *The Social Reproduction of Organization and Culture*, Sociology Volume 2. Sage Publications, London, pp. 1–7.

Himmelstrand, U. (1994), 'Perspectives, Controversies and Dilemmas in the Study of African Development', in Himmelstrand, U. *et al* (eds), *African Perspectives on Development*, James Curry Publication, London, pp 16–36.

Himmelstrand, U. (1995), 'Normative Rationality and Materiality, Rebuilding Sociology after Parsons and Marx, An Autobiographical Perspective', *A Seminal Paper Presented in Madrid,* Autumn.

Hirsh, R. F. (1986), 'How Success Short-Circuits the Future', *Harvard Business Review,* March–April No. 2, pp. 72–86.

Hiskett, M. (1957), 'The Kano Chronicle', *Journal of the Royal Asiatic Society,* Part 1.

Hiskett, M. (1973), *The Sword of Truth: The Life and Times of Usman Dan Fodio*, New Oxford Press, New York.

Hiskett, M (1984), *The Development of Islam in West Africa*, Longman, London.

Hofstede, G. (1980), 'Motivation, Leadership and Organization: Do American Theories Apply Abroad'? *Organizational Dynamics,* Summer, pp. 42–63.

Hofstede, G. (1980a), *Culture's Consequences: International Differences in Work Related Values*, California Sage, Beverly Hills.

Hogendorn, J. S. (1978), *Nigerian Groundnut Exports: Origins and Early Development*, Ahmadu Bello University Press, Zaria.

Hollos, M. (1991), 'Migration, Education and the Status of Women in Southern Nigeria', *American Anthropologist,* No. 93, pp 852–70.

Hoogvelt, A. (1979), 'Indigenisation and Foreign Capital: Industrialization in Nigeria', *Review of African Political Economy,* January/April, pp. 56–67.

Hopkins, A. G. (1973), *An Economic History of West Africa*, Longman, London.

Hountondji, P. (1983), *African Philosophy: Myth and Reality,* Indiana University Press, Blomington.

Howard, M. (1983), *Profit in Economic Theory,* The Macmillan Press Ltd., London and Basingstoke.

Huemer, L. (1998), *Trust in Business Relations: Economic Logic or Social Interaction*? Borea Press, Sweden.

Hugon, P. (1980), 'Seteur Informal et Petite Production Merchande dans les Villes du Tier Monde', *Revue Tier-monde,* April/June, No. 21.

Hydén, G. (1983), *No Shortcuts to Progress, African Development Management Perspectives*, University of California Press, Bekerley.

Ibrahim, J. (1989), 'The Politics of Religion in Nigeria: The Parameters of the 1978 Crises in Kaduna State', *Review of African Political Economy*, No. 45/46.

Ikein, A. A. (1990), *The Impact of Oil on a Developing Country: The Case of Nigeria*, Praeger, New York, and London.

Ikiara, G. K. (1994), 'Entrepreneurship, Industrialization and the National Bourgeoisie in Africa', in Himmelstrand, U. *et al*, (eds), *African Perspectives on Development*, James Currey Publishers, London, pp. 118–127.

Iliffe, J. (1983), *The Emergence of African Capitalism,* The Macmillan Press Ltd., London.

Imo, C. O. (1995), *Religion and the Unity of the Nigerian Nation,* Alqvist and Wiksell International, Stockholm, Sweden.

Industrial Policy of Nigeria (1988), *Policies, Incentives, Guidelines and* Institutional Framework, A Federal Ministry of Industry Publication, Abuja.

Industrialization of Kano State (1970), 'A Guide to Investors', *Kano State, Ministry of Information Publication*, May, Kano.

Inkeles, A. and Smith, D. H. (1974), *Becoming Modern: Individual Changes in Six Developing Countries*, Harvard University Press, Cambridge, Massachusetts.

Isiogu-Abanihe, U. (1994), 'Demographic Transition in the Context of Africa's Development', in Himmelstrand, U. *et al.* (eds), *African Perspectives on Development, Controversies, Dilemmas and Openings, James* Currey, London, pp. 61–73.

Jaggar, P. (1973), 'Kano City Blacksmiths: Pre-Colonial Distribution, Structure and Organization', *Savanna*, Vol. 2, No.1.

Jick, T. D. (1979), 'Mixing Qualitative and Quantitative Method', *Triangulation Administration Service Quarterly*, Vol. 24. No. 4, pp. 602–14.

John, P. *et al* (1976), *The Social Web; An Introduction to Sociology,* Canfield Press, U.S.A.

Jones, G. and Rose, M. B. (1993), 'Family Capitalism', *Journal of Business History*, October, Vol. 35, No. 4.

Kano State (1995), *Ministry of Commerce and Industry Quarterly Report*, Kano.

Kanter, M. R. (1983), *The Change Masters: Corporate Entrepreneurs at Work*, Unwin Paperbacks, London.

Keesing, R. M. (1974), Theories of Culture, *Annual Review of anthropology,* Vol. 3, pp. 73–97.

Keirstead, B. S. (1961), 'Profit', in Sills, D. L. (ed.), *International Encyclopedia of the Social Sciences*, The Macmillan Free Press, USA, Vol. 12, pp. 547–552.

Kent, C. A. *et al.* (1982), *Encyclopedia of Entrepreneurship*, Prentice Hall, Englewood Cliffs, New Jersey.

Kets-de Vries, M. F. F. (1992), 'The Family Firm: An Owner's Manual', *Insead Working Paper Series*, OB, Vol. 92, No. 03.

Khaldun, I. (1967[1377]), *The Muqaddimah, An Introduction to History,* translated by Rosenthal F. Vol. 3, Princeton University Press, Princeton New Jersey.

Kilby, P. (1961), 'African Labour Productivity Reconsidered', *Economic Journal,* June, Vol. lxxi, pp. 273–91.

Kilby, P. (1965), *African Enterprise in Nigerian Bread Industry*, Stanford University Press, Stanford.

Kilby, P. (1969), *Industrialization in An Open Economy*, Cambridge University Press, London.

Kilmann, R. H. *et al.* (eds), (1985), 'Five Steps for Closing Culture-Gaps', in *Gaining Control of Corporate Culture*, Jossey-Bass, San Francisco, pp. 351–69.

Kinberg, L. (1986), 'What is meant by Zuhd', *In Studia Islamica*, lxi, Paris, pp. 27–44.

Kirzner, I (1960), *The Economic Point of View*, Sheed Andrews and Mcmeel, Paris.

Kirzner, I. M. (1973), *Competition and Entrepreneurship*, University Press, Chicago.

Kirzner, I. M. (1983), 'Entrepreneurs and Entrepreneurial Function: A Commentary', in Rosen, J. (ed.), *Entrepreneurship*, Lexicon Books, USA, pp. 281–91.

Klitgaard, R. (1992), 'Taking Culture into Account: From 'Let's' to 'How', in Serageldin, I. and Taborrof, J. (eds), *Culture and Development in Africa*, UNESCO Publications, Washington DC, pp. 75–120.

Knowles, J. C. and Anker R. (1981), 'Determinants of Internal Migration Descriptive Analysis', *Population Report Series*, John Hopkins University, M. No. 7.

Koll, M. (1969), *Craft and Cooperation in Western Nigeria: A Sociological Contribution to Indigenous Economics*, Arnold-Berstraesser-Institute Publication, Germany.

Kreps, D. M. (1990), 'Corporate Culture and Economic Theory', in Alt, J. and Shepsle, K. (eds), *Perspectives on Positive Political Economy*, Cambridge University Press, New York, pp. 90–143.

Krokfors, C. (1995), 'Poverty, Environmental Stress and Culture as Factors in African Migration', in Baker, J. and Aina, T. (eds), *The Migration Experience in Africa*, Gotab, Sweden, pp. 54–64.

Kunkel, J. H. (1971), 'Values and Behaviour in Economic Development', in Kilby, P. (ed.), (1969), *Entrepreneurship and Economic Development,* Macmillan Press, New York, pp. 151–180.

La Barre, W. (1964), 'Para-Linguistics, Kinesics, And Cultural Anthropology', in Thomas, A. *et al.* (eds), *Approaches in Semantics*, Hague.

Lachmann, L. M. (1956), *Capitalism and its Structure*, Buttler and Tanner Press Ltd., London.

Lagerspetz, O. (1992a), 'Legitimacy and Trust', *Philosophical Investigations*, Vol. 15, No. 1, pp. 1–21.

Lander (1980), 1830 *Records of Captain Clapperton's Last Expedition to Africa.* 2 Vols, Colburn & Bentley, London.

Laurent, A. (1983), 'The Cultural Diversity of Western Conceptions of Management', *International Studies of Management and Organization*, Spring-Summer, Vol. xiii, 1–2.

Lazarsfeld, P. F. (1972 [1935]), 'The Art of Asking Why', *Qualitative Analysis*, Allyn and Bacon, Boston.

Lazonic, W. (1991), *Business Organization and the Myth of the Market Economy,* Cambridge University Publications, England.

Leibenstein, H. (1975), 'Aspects of the X-Efficiency Theory of the Firm', *Bell Journal of Economics*, Vol. 6, No. 2.

Leoyd, G. E. R. (eds) (1976), 'Views on Time in Greek Thought', in Gardet, L. *et al. Culture and Time*, The Unesco Press, Paris, pp. 117–48.

Lerner, D. and Pevsner, L. W. (1964) *The Passing of Traditional Society*, The Free Press, New York.

Lester, R. A. (1957), 'Reflections in Collective Bargaining in Britain and Sweden', *Industrial and Labour Relations Review,* April.

Levine, C. (1981), 'Resources Scarcity and Reform Model: The Management of Retrenchment in Cincinnati and Oakland', *Public Administration Review,* Vol. 4, No. 6 pp. 619–628.

Levine, R. A. (1966), *Dreams and Deeds; Achievement and Motives*, University Press, Chicago.

Lewin, K. (1935), *A Dynamic Theory of Personality: Selected Papers*, Mcgraw-Hill, New York.

Lewis, J. D. and A. Weigert (1985), 'Trust as a Social Reality', *Social Forces*, Vol. 63, No. 4, pp. 967–985.

Leys, C. (1965) 'What is the Problem About Corruption', *Journal of Modern African Studies*, Vol. 3, No. 2, pp. 215–30.

Liedholm, C., Chuta, E. (1976), *The Economics of Rural and Urban Small-Scale Industries in Sierra Leone*, Michigan State University Press, AREP, No. 14.

Light, I. (1972), *Ethnic Enterprises in America*, University of California Press, Berkeley.

Liles, P. R. (1974), *New Business Ventures and the Entrepreneur*, Richard D. and Irwin Inc., Momewood, Illinois.

Litvak, I. A. and Maule, C. J. (1973), 'Some Characteristics of Successful Technical Entrepreneurs in Canada', *IEEE Transactions on Engineering Management,* Vol. 20, No. 3, pp. 62–68.

Loeweistein, G. and Mather, J. (1990), 'Dynamic Process in Risk Perception', *Journal of Risk and Uncertainty*, Kluwer Academic Publishers, pp. 155–175.

Lovejoy, P. E. (1973), *The Hausa Kola Trade, 1000–1900,* Ph.D. Dissertation, University of Wisconsin, Wisconsin.

Lubeck, P. M. (1983), 'Industrial Labour in Kano: Historical Origins, Social Characteristics and Sources of Differentiation' in Barkindo, B. M (ed.), *Studies in the History of Kano*, Heinemann Educational Books (Nigeria), Ltd., Ibadan, pp. 147–169.

Lubeck, P. M. (1986), *Islam and Urban Labour in Nigeria; The Making of A Muslim Working Class,* Cambridge University Press, London.

Luhmann, N. (1979), *Trust and Power*, John Wiley and Sons, Avon.

Macdonald, S. (1994), 'Selling British Culture in Europe', In Sjögren, A. and Janson, L. (eds), *Culture and Management in a Changing Europe*, Institute of International Business, Stockholm.

MacMillan, I. C. (1983), 'The Politics of New Venture Management', *Harvard Business Review*, Nov–Dec., pp. 8–12.

Madhock, A (1995), 'Opportunism and Trust in Joint Venture Relationships: An Exploratory Study and a Model', *Scandinavian Journal of Management*, Vol. 11, No.1, pp. 57–74.

Makinwa, P. K. (1981), *Internal Migration and Rural Development in Nigeria,* Heinemann Education Books Nigeria Ltd., Ibadan.

Malik, I. A. (1989), *Al-Muwatta of Imam Malik Ibn Anas: The First Formulation* of Islamic Law, translated by Bewley, A. A. Kegan Paul International, London.

MAN (1994), *'Half Yearly Economic Review'*, January–June, No. 15, Lagos.

MAN (1994), *'Half Yearly Economic Review'*, July–December No. 16, Lagos.

Marc, A. (1992), 'Community Participation in the Conservation of Cultural Heritage', in Serageldin, I. and Taborrof, J (eds), *Culture and Development in Africa*, UNESCO Publications, Washington DC, pp. 255–268.

Marris, P. (1968), 'Social Barriers in African Entrepreneurship', *Journal of Development Studies*, October-Vol. 5, pp. 29–38.

Marris, P. and Somerset, A. (1971), *African Businessmen*, Routledge and Keagan Paul, London.

Marsh, A. (1979), *A Concise Encyclopedia of Industrial Relations,* Gower Press, first edition, USA.

Marx, K. (1969), *Theories of Surplus Value,* Vols. 3, Progress Publishers, Moscow.

Marx, K. (1971 [1844]), 'Wages of Labour', first in the Series of the *Economic and Philosophical Manuscripts,* April/August, Progress Publishers, Moscow.

Marx, K. (1973), *Grundrisse*, Translated by Nicolau, M., Vintage Press, New York.

Marx, K. and Engels, F. (1923 [1848]), *Manifesto of the Communist Party,* translated by More, S. International Publishers, New York.

Maslow, A. (1954), *Motivation and Personality,* Harper and Row, New York.

Mason, E. S. (1989), *The Corporation of Modern Society,* Harvard Press, Cambridge Mass.

Mathews J. and Mathews L. (1983), *One Billion, A China Chronicle*, Ballantine Books, New York.

Maududi, S. A. A. (1983), *Jihad in Islam*, Holy Koran Publishing House, Beirut.

Maut, R. K. and Sharaf, H. A. (1974), 'The Philosophy of Auditing', *American Accounting Association*, USA.

Mazrui, A. A. (1992), 'Development in Multicultural Context: Trends and Tensions', in Serageldin, I. and Taborrof, J. (eds), *Culture and Development in Africa*, UNESCO Publications, Washington DC, pp. 127–136.

Mbiti, J. S. (1971), *African Religions and Philosophy,* Heinemann Publications, London, Ibadan and Nairobi.

McAllister, D. J. (1995), 'Affect and Cognition Based Trust as a Foundation For Interpersonal Cooperation in Organizations', *Academy of Management Journal*, Vol. 38, No.1, pp. 24–59.

McClelland, D. (1961), *The Achieving Society,* Princeton Press, New York.

McGregor, D. M. (1960), *The Human Side of Enterprise,* McGraw Hill, New York.

Mendez-Rivero, D. (1991), *Informalization of the Venezuelan Labour Force,* ISS, The Hague.

Mensah, E. (1982), 'Methodological Problems in the Study of Corruption', in Odekunle, F. (ed.), *Nigeria: Corruption in Development,* University of Ibadan Press, Ibadan, pp. 49–58.

Merton, R. K. *et al* (1956), The Focused Interview, *A Manual of Problems and Procedure*, Glencoe Free Press, Illinois.

Merton, R. K. (1957), *Social Theory and Social Structure,* Glencoe Free Press, Illinoi.

Mescom, M. H. (1981), *Individual and Organizational Effectiveness,* Harper and Row, New York.

Mescom, M. H. *et al* (1985), *Management: Individual and Organizational Effectiveness*, Second Edition, Harper and Row, New York.

Meyer, J. W. *et al* (1977), 'Institutionalized Organizations; Formal Structure, Myth and Ceremony', *American Journal of Sociology,* Vol., 83, No. 2.

Miles, R. H. (1980), *Macro Organizational Behaviour,* Harvard Press, California.
Milgrom, P. R. and J. Roberts (1992), *Economics, Organizations and Management,* Prentice Hall, Englewood Cliffs.
Ministry of Labour (1961), *Industrial Relations Handbook*, U.K.
Mintztberg, H. (1989), *Inside Our Strange World of Organizations,* The Free Press, A division of Macmillan, Inc., New York.
Mockler, R. J. (1972), *The Management Control Process*, Prentice Hall Publications, Englewood Cliffs, New Jersey.
Mohammed, H. D. (1985), 'Women in Nigerian History: Examples from Borno Empire, Nupeland and Igboland, in *Women in Nigeria Today,* (eds), Zed Books Ltd., London, pp. 45–51.
Montagu, A. (1968), 'The Concept of Primitive and Related Anthropological Terms: A study in the Systematics of Confusion', in Montagu, A. (ed.), *The Concept of the Primitive*, The Free Press, New York, pp. 148–168.
Moody, J (1967), 'Staudinger: An Early European Traveller to Kano', *Kano Studies*, Vol. 3, pp. 38–53.
Moran, R. T. and Harris, P. R. (1981), *Managing Cultural Synergy,* Gulf Publishing Company, Houston Texas.
Morgan, G. *et al* (1986), *Images of Organizations*, Sage Publications, Beverly Hills.
Morgan, G. and Smircich, L. (1980), 'The Case for Qualitative Research', *Academy of Management Review,* Vol. 5 No. 4, pp. 491–500.
Mortimore, M. (1989), *Adapting To Drought: Farmers Famines and Desertification in West Africa*, Cambridge University Press: Great Britain.
Mortimore, M. and Wilson, J. (1965), 'Land and People in the Kano Closed-Settled Zone', *Seminal Paper, Department of Geography*, Ahmadu Bello University, Zaria.
Moser, C. (1976), 'Informal Studies or Petty Commodity Production: Dualism or Dependence in Urban Development', *World Development,* September–October, Vol 6.
Mosley, A. G. (1995), *African Philosophy,* Prentice Hall Inc., Englewood Cliffs, New Jersey.
Moughtin, J. C. (1986), *Hausa Architecture*, Lilian Barber Press, New York.
Mustapha, R. A. (1985), 'The Blocked Capitalist Thesis and the Political Economy of Kano', *Nigerian Journal of Political Science*, Vol. 4, Nos. 1 and 2, pp. 24–52.
Myrdal, A. (1968), *Asian Drama: An Enquiry into the Poverty of Nations*, Penguin Books, Great Britain.
Nafziger, W. E. (1967), 'Nigerian Entrepreneurship: A Study of Indigenous Business', Ph.D. Dissertation, Urbana.
Natrass, N. (1984), 'Street Trading in Transkei: A Struggle Against Poverty, Persecution and Prosecution, *Development Studies Working Paper 7*, University Durban, South Africa.
Newman, P. and Roxana (1979), *Modern Hausa-English Dictionary: Sabon Kamus Na Hausa Zuwa Turanci*, Oxford University Press, Ibadan, Zaria.
Newman, W. H. (1975), *Constructive Control. Design and Use of Control Systems*, Englewood Cliffs, Prentice Hall, New Jersey.
Nigerian Constitution (1979), *Federal Ministry of Information,* Lagos.
North, D. C. (1996), *Institutions, Institutional Change and Economic Performance*, Cambridge University Press, New York.
North, D. C. and Thomas, R. P. (1973), *The Rise of the Western World*, Cambridge University Press, Cambridge.

Nyang, S. S. (1992), 'The Cultural Consequences of Development in Africa', in Serageldin, I and Taborrof, J. (eds), *Culture and Development in Africa,* UNESCO Publications, Washington DC, pp. 429–446.

Nye, J. S. (1968), 'Corruption in Development: A Cost-Benefit Analysis', *Development Digest*, Vol. 4, pp. 106–118.

Obregon, Q. (1974), 'The Marginalized Pole of the Economy and the Marginalized Labour Force', *Economy and Society,* Vol. 3.

Odekunle, F. (ed.), (1982), *Nigeria: Corruption in Development,* University of Ibadan Press, Ibadan.

Ogunbanjo, C. O. (1983), *Nigeria's Economic Development; Which Way?,* Nash Publications, London.

Ogunniyi, O. (1991), *Nigerian Labour and Employment Law in Perspective*, Folio Publishers, Lagos.

Okonjo, K. (1980), 'The Role of Igbo Women in Rural Economy in Bendel State of Nigeria', in Omu F. I. A. *et al.* (eds), *Proceedings*, Vol. II, pp. 933–949.

Oloyede, O. (1991), *Coping Under Recession, Workers in a Nigerian Factory,* Ph.D. Dissertation, University of Uppsala, Alqvist and Wiksell International, Stockholm, Sweden.

Olson, M. (1965), *The Logic of Collective Action,* University Press, Harvard.

Olukoshi, A. (1985), 'Some Remarks on the Role of The Levantine Bourgeoisie in the Capitalist Industrialization of Kano', *Nigerian Journal of Political Science*, Vol 4, Nos. 1 and 2.

Olukoshi, A. (1986), A Review of the Development of Capitalism in Northern Nigeria, by Shenton, R., *Review of African Political Economy,* No. 37.

Olusanya, P. O. (1969), *Socioeconomic Aspect of Rural Urban Migration in Western Nigeria*, Nigerian Institute of Social and Economic Research (NISER), Ibadan.

Olzak, S. (1983), 'Contemporary Ethnic Mobilization', *Annual Review of Sociology,* No. 9, pp. 355–374.

Onoge, O. (1982), 'Corruption and the Nigerian Social System: Notes for a Materialist Ethnography', in Odekunle, F. (ed.), *Nigeria: Corruption in Development,* University of Ibadan Press, Ibadan, pp. 20–26.

Opsahl, R. L. and Dunette, M. D. (1966), 'The Role of Financial Compensation in Industrial Motivation', *in Psychological Bulletin*, Washington D.C., pp. 94–118.

Oshagbemi, T. A. (1983), *Small Business Management in Nigeria*, Longmann Publications, London and Lagos.

Osuntogun, A. (1976), 'Some Causes and Implications for Socioeconomic Development of City Ward Migration of Rural Youth: A Western Nigerian Case Study', in Adepoju, A. (ed.), *Internal Migration in Nigeria,* University of Ife Press, Ife.

Otobo, D. (1987), 'Bureaucratic Elites and the Public Sector, Wage Bargaining in Nigeria', in Otobo, D, and Omole, M, (eds), *Readings in Industrial Relations in Nigeria,* Malt House Publishing Ltd., UK, pp. 45–71.

Paden, J. N (1973), *Religion and Political Culture in Kano*, University of California Press, USA.

Peace, A. (1979), *Choice, Class and Conflict: A Study of Southern Nigerian Factory Workers*, The Harvester Press, Great Britain.

Peacock, J. L (1972), *Muslim Puritans: Reformist Psychology in Southeast Asian Islam*, University of California Press, USA.

Peil, M. *et al.* (1982), *Social Science Research Methods*, Evans Publications, London.

Peil, M. and Sada P. O. (1984), *African Urban Society*, John Willey and Sons, Chichester.

Pfeffer, J. (1981), Management of Symbolic Action: The Creation and Maintenance of Organizational Paradigms, in Cummings, L. L. and Staw, B. M. (eds), *Research in Organizational Behaviour*, CT: JAI Press, Greenwich, Vol. 3.

Phillips, A. (1977), 'The Concept of Development', *Review of African Political Economy*, January/April, pp. 7–20.

Pickles, A. R. and Farrell, P. N. (1986), 'An Analysis of Entrepreneurial Behaviour from Male Work Histories', *Regional Studies*, Vol. 21, No. 5, pp. 425–444.

Piore, M. J. (1979), *Birds of Passage: Migrant Labour and Industrial Societies*, Cambridge University Press, London.

Piore, M. J. and Sabel, C. F. (1984), *The Second Industrial Divide: Possibilities For Prosperity*, Basic Books, New York.

Pittin, R. (1983), 'Houses of Women: A Focus on Alternative Life Styles in Katsina City', in Oppong, C. (ed.) *Female and Male in West Africa*, George Allen and Unwin, London, pp.291–302.

Pittin, R. (1984), 'Migration of Women in Nigeria: The Hausa Case', *International Migration Review*, No. 18, pp. 1293–1314.

Pokrant, R. (1982), 'The Tailors of Kano City', in Goody E. (ed.), *From Craft to Industry*, Cambridge University Press, Cambridge.

Porter, M. E. (1980), *Competitive Strategy*, Free Press, New York.

Poughkeepsie, N. Y. (1969), 'Length of Life of Enterprises Established Between 1844 and 1926', *Duns Review*.

Powell, W. (1982), *Saudi Arabia and its Royal Family*, Lyle Stuart, Secausus, New Jersey.

Raiffa H. (1968), *Decision Analysis*, Addison Wesley, Reading M. A.

Ranis, G. and Fei, J. C. H. (1961), 'A Theory of Economic Development', *American Economic Review*, September, Vol. 51, No 4.

Ray, C. A. (1986) 'Corporate Culture, the Last Frontier of Control', *Journal of Management Studies*, Vol. 23, No. 3.

Ray, D. (1988), 'The Role of Entrepreneurship in Economic Development', in Leibenstein, H. and Ray, D. (eds), Entrepreneurship and Economic Development, *Journal Of Development Planning*, UN, New York, No. 18.

Redding, G. R. (1990), *The Spirit of Chinese Capitalism*, Berlin.

Redding, G. S. (1986), 'Developing Management With Management Development', *Administrative Science Quarterly*, Vol. 5, No. 1, pp. 89–96.

Ricardo, D. (1986 [1955]), 'On the Principles of Political Economy of Taxation', Vol. 1, *Works and Correspondence of David Ricardo*, (ed.), by Sraffa, P. in Collaboration with Dobb, M., Cambridge University Press, Cambridge.

Ritchie, B. and Marshall, D (1993), *Business Risk Management*, Alden Press Oxford, Britain.

Robinson, J. (1961), *The Economics of Imperfect Competition*, London.

Robinson, R. (1965), *Industrialization in Developing Countries*, Cambridge University Press, England.

Rodinson, M. (1980), *Islam and Capitalism*, a reprinted edition, Penguin Series, Great Britain.

Rodney, W (1972), *How Europe Underdeveloped Africa*, Bogle-L'Ouverture and Tanzania Publishing House, London and Dar-es-Salam.

Ross, A. M. (1966), *Trade Union Wage Policy*, University of California Press, Berkeley and Los Angeles.

Rowe, M. P. (1971), *Indigenous Industrial Entrepreneurship in Lagos Nigeria*, Columbia University Press, USA.

Sabel, C. F. (1993), 'Studied Trust: Building New Forms of Co-operation in a Volatile Economy', *Human Relations*, Vol. 46, No. 9, pp. 1133–70.

Sackmann, S. A. (1991), *Cultural Knowledge in Organizations: Exploring the Collective Mind*, Sage Publications, Inc., USA.

Salim, A. S. (1992), 'Opening Remarks', in Serageldin, I. and Taborrof, J. (eds) *Culture and Development in Africa*, UNESCO Publications, Washington DC, pp. 9–13.

Sanbrook, R. and Barker, J. (1985), *The Politics of Africa's Economic Stagnation*, Cambridge University Press.

Sanner, L. (1997), *Trust Between Entrepreneurs and External Actors: Sensemaking in Organizing New Business Ventures*, Doctoral Thesis, The Department of Business Studies, Uppsala University, Sweden.

Sathe, V. (1983), 'The Controller's Role in Management', *Organizational Dynamics*, Winter, Vol. 11, No. 3.

Schatz, P. S. (1977), *Nigerian Capitalism*, University Press, California.

Schildkrout, E. (1981), 'The Employment of Children in Kano, Nigeria', in Gerry Rodger, G. and Stanley, G. (eds), *Childwork Poverty and Underdevelopment*, International Labour Office, Geneva, pp. 195–223.

Schildkrout, E. (1983), Dependence and Autonomy: The Economic Activities of Secluded in Kano', in Oppong, C. (ed.), *Female and Male in West Africa*, George Allen and Unwin, London, pp. 106–26.

Schildkrout, E (1986), 'Entrepreneurial Activities of Women and Children Among the Islamic Hausa of Northern Nigeria', in Greenfield, S. M. and Strickon, A. (eds), *Entrepreneurship and Social Change, Monographs in Economic Anthropology*, No. 2. University Press of America, pp. 195–223.

Schneider, B. and Rentsch, J. (1988), 'Managing Climate and Cultures: A Futures Perspective', in Hage, J. (ed.), *Futures of Organizations*, Lexington Books, Lexington M. A.

Schoemaker, P. J. H. (1993), 'Determinants of Risk-Taking and Uncertainty Behavioural and Economic Views', *Journal of Risk and Uncertainty*, Kluwer Academic Publishers, No. 6, pp. 49–73.

Schumpeter, J. A. (1934), 'The Theory of Economic Development: An Enquiry into Profit, Capital, Credit, Interest and Business Cycle', *Harvard Economic Studies*, Cambridge Mass, Vol. 46.

Schumpeter, J. A. (1962), *Capitalism, Socialism and Democracy*, Harper Torch Books, New York.

Scott, R. W. (1986), 'The Sociology of Organizations', in Himmelstrand, U. (ed.), *The Social Reproduction Of Organization and Culture*, Sage Publications, London, pp. 38–58.

Seers, D. (1964), *The Meaning of Development*, International Development Review, Vol. 11, No.4.

Segall, D. T. (1963), 'Cultural Differences in the Perception of Geometric Illusions', *Science*, Vol. 139, pp. 769–71.

Selznick, P. (1957), *Leadership in Administration; A Sociological Interpretation*, Row and Peterson, New York.

Serageldin, I. (1992), 'The Challenge of a Holistic Vision: Culture, Empowerment and the Development Paradigm', in Serageldin, I. and Taborrof, J. (eds), *Culture and Development in Africa*, UNESCO Publications, Washington DC, pp. 15–32.

Sethuraman, S. (1981), *The Urban Informal Sector in Developing Countries: Employment, Poverty and Environment*, ILO Publication, Geneva.

Shätzl, L. (1973), *Industrialization in Nigeria: A Spacial Analysis,* Germany.

Shea, P. J. (1974), 'Economies of Scale: The Dying Industry of Pre-colonial Kano', *Kano Studies, New Series,* Vol. 1, No. 2.

Shea, P. J. (1983), 'Approaching the Study of Production in Rural Kano', in Barkindo, B. M. (ed.), *Studies in the History of Kano*, Heinemann, Educational Books, Ibadan, pp. 93–115.

Shenton, R. (1986), *Development of Capitalism in Northern Nigeria*, James Curry, London.

Sherif, M. H. (1987), 'What is Hijab'? *Muslim World*, (July–October) Vol. 7, Nos. 3–4, pp. 151–63.

Silin, R. H. (1976), *Leadership and Values: The organization of Large Scale Taiwanese Enterprises*, Cambridge Mass, Harvard.

Simms, R. (1981), 'The African Woman as Entrepreneur: Problems and Perspectives on Their Roles', in Steady, F. (ed.), *The Black Woman Cross-Culturally*, Schenkman, Cambridge, Mass, pp. 141–68.

Simon, J. L. (1982), 'An Answer to Malthus', Interview by Buckley, W. *Population and Development Review*, Vol. 8, No. 1.

Sitkin, S. B and N. L. Roth (1993), 'Explaining the Limited Effectiveness of Legalistic 'Remedies' for Trust/Distrust', *Organization Science*, August, Vol. 4, No. 3, pp. 367–392.

Smircich, L. (1983a), 'Concepts of Culture and Organizational Analysis', *Administrative Science Quarterly*, Vol. 28, No. 3, pp. 339–58.

Smircich, L. (1983c), 'Organizations as Shared Meaning', in Pondy, L. R. *et al.* (eds), *Organizational Symbolism*, JAI Press, Greenwich.

Smith, A. ([1776, 1789] 1981), *An Enquiry into the Nature and Causes of Wealth of Nations*, Vol 11 (ed.), by Campbell, R. H. *et al.*, Liberty Classics, Indianapolis.

Smith, M. G. (1955), 'The Economy of the Hausa Communities of Zaria', *Colonial Research Studies*, HMSO, London, No. 16.

Smith, M. G. (1962), 'Exchange and Marketing Among the Hausa', in Bohanan, P. and Dalton, G. (eds), *Markets in Africa*, Evans Press, Britain.

Smith, M. G. (1964), 'The Historical and Cultural Conditions of Political Corruption Among the Hausa, *Comparative Studies in Society and History*, Vol. 6. No. 2, pp. 164–194.

Smith, M. G. (1983), 'The Kano Chronicle as History' in Barkindo, B. M. (ed.), *Studies in the History of Kano*, Heinemann Educational Books, Ibadan, pp. 31–56.

Sokolski, A. (1965), *The Establishment of Manufacture in Nigeria*, Praeger, New York.

Sonubi, O. (1987), 'Trade Unions and their Members', in Otobo, D. and Omole, M. (eds) *Readings in Industrial Relations in Nigeria*, MaltHouse Publishing Ltd., UK, pp. 28–35.

Soyinka, W. (1992), 'Culture, Memory and Development', in Serageldin, I. and Taborrof, J. (eds), *Culture and Development in Africa*, UNESCO Publications, Washington DC, pp. 201–18.

Sraffa, P. ([1960] 1973), *Production of Commodities by Means of Commodities: Prelude to a Critique of Economic Theory*, Cambridge University Press, Cambridge.

Steed, C. (1995), 'The Islamic Heritage of Nigeria', in Steed, C. and Westlund, D. (eds), *The Triple Religious Heritage of Nigeria: Three Essays*, Teologiska Institutionen, Uppsala Universitet, Sweden, pp. 67–83.

Steers, R. M. (1981), *Introduction to Organizational Behaviour*, Scott, Foresman and Company, Illinoi.

Stenmar, M. (1994), *Rationality in Science, Religion and Everyday Life: A Critical Evaluation of Four Models of Rationality*, Doctoral Dissertation, Department of Theology, Uppsala University, Sweden.

Stoner, J. A. F. and Freeman, R. E. (1989), *Management*, Forth Edition, Prentice Hall, Englewood Cliffs, New Jersey.

Stump, K. W. (1984), *A Crucial Half-Century of Religion*, London.

Sulivan, J. O. (1987), 'Memories of Development', in Berger, P. L. (ed.), *Capitalism and Equality in the Third World*, Hamilton Press, London.

Tahir, I. (1975), 'Scholars, Sufis, Saints and Capitalists in Kano 1904–1974; A Pattern of a Bourgeois Revolution in an Islamic Society', Ph.D. Dissertation, Cambridge University.

Takamiya, S. (1 97), 'Group Decision Making in Japanese Management', *International Studies of Management and Organization*, Summer, Vol. 2, No. 2.

Teriba, O. *et al.* (eds), (1977), *Industrial Development in Nigeria, Patterns, Problems and Prospects*, University of Ibadan Press, Ibadan.

Thorne, J. R. and Ball, J. G. (1981), 'Entrepreneurs and their Companies', in Vesper, K. H. (ed.), *Frontiers of Entrepreneurship Research*, Babson College,Wellsley, Mass.

Thornhill, W. T. (1989), *Effective Risk Management for Financial Organizations,* Bank Administration Institute, Illinois.

Tichy, N. and Fombrum C. (1979), 'Network Analysis in Organizational Setting', *Human Relations*, Vol. 32, No. 11, pp. 923–965.

Tiffin, S. and Osotimehin, S. O. A. (1988) 'Technical Entrepreneurship and Entrepreneurial Innovation in Nigeria', *Journal of Development Planning*, No. 18, pp. 195–220.

Tiger, L. (1969), *Men in Groups*, Random House, New York.

Tilman, R. O. (1963), *The Nigerian Political Scene*, Drake University Press, Carolina.

Todaro, M. P. (1969), 'A Model of Labour Migration and Urban Unemployment in Less Developed Countries', *American Economic Review*, Vol. 59, No. 1.

Tokman, V. (1978), 'An Exploration into the Nature of the Informal/Formal Sector Relationships', *World Development,* September–October, No. 6.

Tong, R. M. (1978), 'Synthesis of Fuzzy Models for Industrial Process: Some Recent Results', International Journal of General System, No. 4, pp. 143–62.

Trimingham, S. J. (1971), *The Sufi Orders in Islam*, Oxford University Press, Great Britain.

Turner, S. B. (1974), *Max Weber and Islam*, Routledge and Keagan Paul, London.

Ubeku, A. K. (1983), *Industrial Relations in Developing Countries:* The Case of Nigeria, Macmillan Press, London.

Udehn, L. (1996), *The Limits of Public Choice, A Sociological Critique of the Economic Theory of Politics*, Routledge Publications, London and New York.

Ulin, R. C. (1988), *Understanding Cultures-Perspectives in Anthropological Social Theory*, University of Texas Press, Texas.

Van Dijk, P. M. (1994), 'New Competition and Flexible Specialization in Indonesia and Burkina Faso', in Perdersen, P. O. *et al.* (eds), *Flexible Specialization: The Dynamics of Small-Scale Industries in the South*, Intermediate Technology Publication, London, pp. 55–70.

Van-Roy, E. (1970), 'The Theory of Corruption', *Economic Development and Cultural Change*, October, Vol. xix. 1.

Vernom, P. E. (1965b), 'Abilities Factors and Environmental Influences', *American Psychologist,* Vol. 20.

Vesper, K. H. (1980), *New Venture Strategies,* Prentice Hall, Englewood Cliffs, New Jersey.

Vickery, B. G. (1973), *Principles and Practices of Book Keeping and Accounts*, Donnington Press.

Viddich, A. J. (1955), 'Participant Observation and the Collection and Interpretation of Data', *American Journal of Sociology*, Vol. 60, pp. 354–60.

Vroom, V. (1960), *Some Personality Determinants of the Effects of Participation*, Prentice Hall, Englewwod Cliffs New Jersey.

Waldinger, R. *et al.* (1990), *Ethnic Entrepreneurs: Immigrant Business in Industrial Societies*, Sage Series, London and New Delhi, Vol. 1.

Walras, L. (1984 [1874,1926]), *Elements of Pure Economics, Or the Theory of Social Wealth*, Orion Editions, Philadelphia.

Ward, M. (1976), *The Measurement of Capital: The Methodology of Capital Stock Estimates in OECD Countries*, OECD Publication, Paris.

Washington Post (1996), 16 October.

Waterman, R. J. (1987), 'The Renewal Factor', *Business Week*, September 14.

Watts, M. (1983), *Silent Voices*, University of California Press, Berkeley.

Watts, M. J. (1987), 'Peasantry, Merchant Capital, and the Colonial State: Class in Northern Nigeria, 1900–1945', in Lubeck, P. M. (ed.), *The African Bourgeoisie: Capitalist Development in Nigeria, Kenya and the Ivory Coast,* Lynne Rienner Publishers, Boulder, Colorado, pp. 59–96.

Web, E. J. *et al.* (1966), *Unobtrusive Measures: Nonreactive Research in Social Sciences*, Rand McNally, Chicago.

Weber, M (1947), *The Theory of Social and Economic Organization*, Translated by Parsons, T. and Henderson, A. M., Oxford University Press, New York.

Weber, M. (1974), *From Max Weber*, Translated with an Introduction by Gerth, H. H. and Mills, C. W., Routledge and Keagan Paul Ltd., London and Boston.

Weber, M. (1976), *The Protestant Ethic and the Spirit of Capitalism*, Translated by Parsons, T. and Giddens, A., Allen and Unwin, London and Boston, London.

Weber, M. (1978), *Economy and Society*, Vol. 1., University of California Press, Berkeley and Los Angeles.

Weber, M. (1980), *General Economic History*, translated by Knight, F. H., Allen and Unwin, London.

Weeks, J. A. (1968), 'A Comment on Peter Kilby: Industrial Relations and Wage Determination', *Journal of Developing Areas*, Vol. 3, pp. 7–18.

Werlin, H. H. (1972), 'The Roots of Corruption: The Ghanaian Enquiry', *Journal of Modern African Studies*, Vol. 10, No. 2.

Westley, F. and Jaeger, A. (1985), 'An Examination of Organizational Culture: How it is

Linked to Performance?', *Seminal Paper, Faculty of Management,* Mcgill University, Montreal.

Wildavsky, A. (1992), 'How Cultural Theory Can Contribute to Understanding and Promoting Democracy, Science, and Development', in Serageldin, I. and Taboroff, J. (eds), *Culture and Development in Africa*, World Bank Publication, Washington D.C., pp. 137–64.

Wilken, P. H. (1979), *Entrepreneurship: A Comparative and Historical Study,* N J I, Alex Publishing Company, Norwood.

Windmuller, J. P. *et al* (1987), *Collective Bargaining in Industrialized Market Economies: An Reappraisal*, ILO, Geneva.

Wood, A. (1975), *A theory of Profit*, Cambridge University Press, Britain.

Woodruff, A, Alexander, T. (1958), *Success and Failure in Small Manufacturing*, Pittsburgh.

Woodward, A. E. *et al.* (1994), *Municipal Entrepreneurs and Energy Policy: A five Nation Study of Politics, Innovation and Social Change*, Gordon and Breach, USA.

World Bank (1978), 'World Development Report', Washington D. C.

World Bank (1989), *Sub-Saharan Africa, From Crises to Sustainable Growth*, Washington D.C.

World Bank (1994), *Nigeria: Structural Adjustment Program, Policies, Implementation and Impact*, May, Report No. 130553-UNI.

Worsley, P. (1984), *The Three Worlds: Culture and World Development*, Chicago Press, USA.

Wraith, R. and Simpkins, E. (1963), *Corruption in Developing Countries,* London.

Yesufu, T. M. (1962), *An Introduction to Industrial Relations in Nigeria*, Oxford University Press, England.

Yesufu, T. M. (1967), 'Consumption Patterns and Living Conditions of Workers in the Nigerian Tobacco Company Limited', *Seminal Paper*, University of Ibadan.

Yusuf, A. B. (1975), 'Capital Formation and Management Among the Hausa Traders of Northern Nigeria', *Africa,* xlx, pp. 167–82.

Young, L. C. (1992), *The Role of Trust in Interorganizational Relationships in Marketing Channels*, University of New South Wales, Doctoral Thesis, Sydney.

Zaharadeen, M. S. (1983), 'The Place of Mosques in the History of Kano', in Barkindo, B. M. (ed.), *Studies in the History of Kano*, Heinemann Educational Books Nigeria Ltd., Ibadan, pp. 57–66.

Zetterberg, H. L. (1965), *On Theory and Verification in Sociology,* Third and Enlarged Edition, Bedminster Press, USA.

Index